THE WRYNECK

Adult Wryneck. June 2018, Traisen, Austria (TH).

THE WRYNECK

Biology, Behaviour, Conservation and Symbolism of *Jynx torquilla*

GERARD GORMAN

PELAGIC PUBLISHING

First published in 2022 by
Pelagic Publishing
PO Box 874
Exeter, EX3 9BR, UK

www.pelagicpublishing.com

The Wryneck: Biology, Behaviour, Conservation and Symbolism of *Jynx torquilla*

British Library Cataloguing in Publication Data
A catalogue record for this book is available
from the British Library

ISBN 978-1-78427-288-3 Pbk
ISBN 978-1-78427-289-0 ePub
ISBN 978-1-78427-290-6 PDF

https://doi.org/10.53061/VHKC7954

Cover photo: Wryneck © Mathias Schaef/McPhoto/ullstein bild via Getty Images

Typeset by BBR Design, Sheffield

Printed in Wales by Cambrian Printers, The Pensord Group

To the memory of my father, Robert, for taking me out into the countryside.
To the memory of my mother, Joyce, for everything.
To my boys, Martin and Dominic, for their patience and tolerance.

Adult Wryneck with a bill full of ants. June 2021, Pest County, Hungary (RP).

Contents

The author among the deadwood, Białowieża Forest, Poland.

About the author

If you have arrived here because you are captivated by woodpeckers, then you probably already own a book or two by Gerard Gorman. Quite simply, Gerard is an authority on the Picidae. He has published numerous papers, notes and an unprecedented six previous books on these fascinating birds: *Woodpeckers of Europe: A Study of the European Picidae* (Coleman 2004), *The Black Woodpecker: A Monograph on* Dryocopus martius (Lynx 2011), *Woodpeckers of the World: The Complete Guide* (Helm 2014), *Woodpecker* (Reaktion 2017), *Spotlight Woodpeckers* (Bloomsbury 2018) and *The Green Woodpecker* (Picus Press 2020). He has been asked many times 'Why woodpeckers?' and his usual answer is 'It was inevitable'. Yet, when pressed, he recounts that he preferred to walk in woodlands rather than sit at sewage farms and rubbish tips when birding in his youth, and that this was probably how his passion for picids began. Subsequently, for the past 30 years he has travelled the world searching for and studying woodpeckers, believing that above all else time in the field is the only way to try to get to know them. In this book Gerard has augmented his many hours watching Wrynecks with comprehensive literature research to create what will surely become the definitive work on the species. He lives in Budapest and is a founder member and current leader of the Hungarian Woodpecker Working Group.

Acknowledgements

An acquaintance suggested that I was obsessed. But, apparently with a wry smile on my face, I told him he was wrong, that I was merely focused. I've observed Wrynecks in the field, in all seasons, on three continents. I've handled them, examined museum specimens, taken photographs, made sound recordings of their vocalisations and placed out nest boxes for them. All that, however, was never going to be enough – this species is far too complex. You think you know a bird, then one day you go out and there it is, behaving in a way you believed it did not do. So, I sought out others with experience of Wrynecks and I am grateful for their encouragement and contributions. In particular, I must thank Nigel Massen and David Hawkins at Pelagic Publishing, who took this project on board and were supportive throughout. Daniel Alder, David Christie, Thomas Hochebner, Chris Kehoe and Peter Powney read sections of my manuscript and improved it considerably. Kyle Turner shared his in-depth knowledge of sounds. Rolland Kern, Antal Klébert and András Schmidt were enthusiastic partners in 'Operation Wryneck' in Hungary. Thanks, too, to Gergely Babocsay (Mátra Museum of the Hungarian Natural History Museum), Hans-Martin Berg (Natural History Museum Vienna), Tibor Fuisz (Hungarian Natural History Museum Budapest) and Stephan Weigl (Biologiezentrum des Oberösterreichischen Landesmuseums, Linz).

The following also helped in various ways: Vasil Ananian, Sveta Ashby, Vaughan Ashby, Imre Bárdos, Leon Berthou, Herbert Boedendorfer, Jean-Michel Bompar, Nik Borrow, Duncan Butchart, Ioana Catalina, Sayam Chowdhury, Josef Chytil, Philippe Collard, Peter Cosgrove, Robert Dowsett, Françoise Dowsett-Lemaire, Martin Dvorak, Upali Ekanayake, Tomasz Figarski, Nicholas Galea, Chris Galvin, Dimiter Georgiev, Luís Gordinho, Paul Harris, Rolf Hennes, Axel Hirschfeld, Erik Hirschfeld, Remco Hofland, Ayuwat Jearwattanakanok, Phil Jones, Łukasz Kajtoch, Ondřej Kauzál, Denis Kitel, Benjamin Knes, Josip Ledinšćak, Li Chung Hoi Tom, Mati Kose, Godfrey McRoberts, Karlis Millers, István Moldován, Silas Olofson, Samuel Pačenovský, David Parnaby, Yoav Perlman, Nikolai Petkov, Tatiana Petrova, Petr Prochazka, Dave Pullan, István

Rottenhoffer, Milan Ružić, Gergő Sári, Georg Schreier, Domen Stanič, Bård Stokke, Dániel Szimuly, Ehsan Talebi, Warwick Tarboton, Norbert Teufelbauer, Antero Topp, Rudi Triebl, Josip Turkalj, Anna Ufimzewa, Luis Mario Arce Velasco, Ray Vella, Hans Winkler, Rick Wright and Árpád Zsoldos.

Images taken by these talented people greatly enhance this book: Zdeněk Abrahámek (ZA), Sveta Ashby (SA), Terry Ayling (TA), Imre Bárdos (IB), Jean-Michel Bompar (JMB), Nik Borrow (NiB), Neil Bowman (NeB), Josef Chytil (JC), José María Fernández-Díaz Formentí (JMFDF), Rob Daw (RD), Chris Galvin (CG), Dimiter Georgiev (DG), Tomáš Grim (TG), Paul Harris (PH), Thomas Hochebner (TH), Antal Klébert (AK), Tibor Pecsics (TP), Rudi Petitjean (RP), Aleš Toman (AT), Elena Ternelli (ET), Ray Vella (RV), Stephan Weigl (SW), the Committee Against Bird Slaughter (CABS). My own images are followed by the initials (GG).

Preface

> Some have two toes in front and two behind, such as the one called the Wryneck (iunx). This bird is a little bigger than the Chaffinch, and it is mottled. The arrangement of its toes is unique ... Its claws are large, like those of Green Woodpecker. It makes a high-pitched noise.
>
> Aristotle, *History of Animals*, 350 BCE.

What, exactly, is a Wryneck? In short, Wrynecks are woodpeckers. That fact itself often surprises people, but these birds are indeed members of the avian family Picidae. There are two species, both in the Old World: the Eurasian Wryneck *Jynx torquilla* and the Red-throated Wryneck *Jynx ruficollis*. This book focuses on the former but I do include a chapter on the latter, which I have observed in South Africa and Uganda, for comparison.

Although they are woodpeckers, these birds are far from typical members of the family. Ornithologists refer to them as 'aberrant' or 'atypical' and birdwatchers often exclaim that they are 'bizarre' or 'weird', as they differ from their 'typical' woodpecker relatives in several ways. For example, they are perching rather than climbing birds, and they do not make their own nest holes in trees, nor bore into them to find insect prey. In fact, they are unable to open up wood for any purpose as they lack most of the key anatomical adaptations to do so, such as a robust bill and rigid tail feathers, which the 'true woodpeckers' have evolved. Of the 11 species in the Picidae that breed in Europe, the Wryneck is the only one that is truly migratory, most heading south each autumn to winter in sub-Saharan Africa and then returning in the spring to breed.

A search through the literature on *Jynx torquilla* will reveal that besides the name Eurasian Wryneck, this species is also referred to in English as Northern Wryneck and often as simply Wryneck. In the interests of brevity, I also use the short version throughout this book, with the exception of a few instances where I write Eurasian Wryneck to make a clear distinction from the Red-throated Wryneck. So, 'Wryneck' – an unusual name? For some perhaps also a little alarming as this word also refers to torticollis, a medical condition in people caused by spasmodic contractions of the neck muscles, which results in the neck twisting and the head inclining to one side. Wrynecks do not suffer from that disorder but will bend their neck and rotate their head awry when threatened. Hence the odd name.

I have been writing this book for several years. There were occasions when I was about to complete it and then I'd see something in the field, or come upon another paper or note on the species that contained something new that meant I had to go back to my manuscript and rephrase a line or two. Writing a book like this can become a never-ending undertaking, but there comes a moment when you simply have to stop. That decision was made easier for me when 2020 arrived and I, like so many others, found myself with more desk-time than I had ever had. That situation, and a prompt by a fellow woodpecker aficionado to 'Write a book about the weirdest woodpecker of all' forced me to get it done. So here is *The Wryneck*, a combination of my own experience with this intriguing bird, the knowledge of some fine friends and as much published research on the species as I could find.

Adult Wryneck. April 2021, Gerecse Hills, Hungary (GG).

This book is for everyone: not only for woodpecker fanatics (and there are a few), but for ornithologists – professional and amateur – citizen naturalists, birdwatchers, and for those who just love to be outdoors observing all wildlife.

Finally, there is another reason. I wanted to bring more attention to a bird that, although seemingly still common in many parts of its range, is actually in decline. Wrynecks have already disappeared from Britain as a breeding bird, jinxed by things we still do not fully understand, and elsewhere people are noticing that numbers have fallen. The Wryneck is an extraordinary creature, one that is synonymous with twists and turns, of which, sadly, I feel there are a few more to come in its world.

Gerard Gorman, November 2021

Chapter 1

Origins and Taxonomy

Woodpeckers are members of the Picidae, a family of birds in the order Piciformes. They are the most widespread and largest single family of the Piciformes, occurring from sea-level to high mountains on every continent apart from Oceania (Australia, New Zealand, Papua and islands east of there) and Antarctica. They are absent also from Madagascar and (perhaps less of a surprise) the Arctic. Studies of genetics (molecular-sequence analysis) and morphology (form and physique) indicate that the woodpeckers' closest relatives are the honeyguides (Indicatoridae) of Africa and Asia. Other Piciformes families are the barbets of Africa (Lybiidae), Asia (Megalaimidae) and South America (Capitonidae), and the South and Middle American toucans (Ramphastidae), puffbirds (Bucconidae) and jacamars (Galbulidae) (Winkler 2015).

Evolution

The evolutionary history of woodpeckers and the relationships between the different families and species are still not fully understood. It is largely agreed, however, that in evolutionary terms woodpeckers are relatively advanced birds. The 'true' woodpeckers, in the subfamily Picinae, have evolved many anatomical features that have developed in line with their morphology (see Chapter 2). These adaptations enable woodpeckers to take advantage of ecological niches that most other birds, and indeed most other vertebrates, cannot exploit. The question of whether these adaptations evolved independently or whether they are a result of reverse evolution (re-adaptation to an earlier, ancestral habitat or environment) is one that remains to be satisfactorily answered (Short 1982). In an evolutionary sense, wrynecks have been considered by some scholars to be the woodpeckers that most resemble

FIGURE 1.1 An adult Wryneck in a rather 'reptilian' pose. June 2013, Kocsér, Hungary (RP).

ancestral members of the family (Benz et al. 2006) and have even been called the most primitive woodpeckers (Goodge 1972). As they lack many of the physical adaptations (in particular those that relate to protecting the brain from impacts and vibrations) that their relatives possess, wrynecks might be regarded as 'proto-woodpeckers'. On the other hand, it may be the case that they are the most advanced in evolutionary terms, being woodpeckers that have shifted from a mainly arboreal existence to a more terrestrial one, feeding on ground-dwelling prey rather than on those invertebrates that live on or in trees. For the moment, the positioning of wrynecks in evolutionary history remains a chicken-and-egg question.

Fossils

The fossil record suggests that the Picidae first evolved in what is today Europe and Asia after diverging from their early relatives in the Piciformes order about 50 million years ago (Sibley and Ahlquist 1990). The true woodpeckers as we know them today are probably quite similar to those that lived in the Pliocene epoch, around 5 million years ago (Winkler 2015). Fossils discovered in Europe reveal that they were already present at the beginning of the Paleogene era (*c.*66 to 23 million years ago), represented by species of the tropical Capitonidae and the now extinct Zygodactylidae families. Fossils of birds that resemble the woodpeckers we see today are scarce, although some have been found in Pliocene epoch deposits in the Carpathian basin (Kessler 2014). Skeletal fossil remains of wrynecks are known from the Pleistocene epoch (commonly called the Ice Age), which was some 2.6 million to 11,700 years ago, for example, from the Lower Pleistocene in Hungary and Romania, the Middle Pleistocene in France, and the Upper Pleistocene in Austria, Croatia, the Czech Republic, France, Germany, Romania and Switzerland (Kessler 2016).

Taxonomy

The classification of woodpeckers is far more complicated than previously believed. The apparent morphological uniformity of woodpeckers conventionally led most taxonomists (and also most birdwatchers) to classify them primarily on plumage features, but today the analyses of molecular data have begun to revise and improve woodpecker taxonomy (Winkler et al. 2014). Currently, the woodpecker family (Picidae) is divided into three subfamilies, the Jynginae (wrynecks), the Picumninae (piculets) and the Picinae (true woodpeckers). DNA sequencing and phylogenetic analyses have confirmed that Jynginae are sister to other woodpeckers, including the Picinae (Du et al. 2020). The Jynginae subfamily contains one genus, *Jynx*, with two species: *Jynx torquilla* (Eurasian Wryneck) and *Jynx ruficollis* (Red-throated Wryneck). The two wrynecks form a superspecies which most likely diverged from the piculets, and from the birds that subsequently evolved into the true woodpeckers, at an early stage in the evolutionary history of the Picidae (Winkler 2015). They have an exclusively Old World distribution, occurring in Europe, Asia and Africa (Gorman 2014). They coincide only in a few areas in Africa where the Eurasian species, in its southernmost wintering quarters, meets the Red-throated, which is resident. For more on the Red-throated Wryneck see Chapter 5.

Jynx torquilla was first described and classified by Carl von Linné (Linnaeus) in 1758, in the tenth edition of his *Systema Naturae*. The type-locality was Sweden. The species is placed in the overall taxonomy of the animal kingdom as follows:

- ***Kingdom:*** Animalia (animals)
- ***Phylum:*** Chordata (vertebrates)
- ***Class:*** Aves (birds)
- ***Order:*** Piciformes (woodpeckers, honeyguides and allies)
- ***Family:*** Picidae (woodpeckers, piculets, wrynecks)
- ***Subfamily:*** Jynginae (wrynecks)
- ***Genus:*** *Jynx* (two members)
- ***Species:*** *Jynx torquilla* (Eurasian Wryneck), *Jynx ruficollis* (Red-throated Wryneck)

The subspecies

The Eurasian Wryneck is polytypic, taxonomists and checklist authors recognising from four to seven subspecies (Vaurie 1959; Winkler et al. 1995; del Hoyo and Collar 2014; Clements et al. 2019; Gill et al. 2020). The nominate *torquilla* breeds across most of continental Europe, the eastern Balkans and eastwards through Russia and the Caucasus into Asia. Most of its European population

FIGURE 1.2 Adult Wryneck nominate subspecies *torquilla*. May 2021, Novo Yankovo, Bulgaria (DG).

FIGURE 1.3 Adult Wryneck subspecies *tschusii*. June 2020, Modena, Italy (ET).

appears to winter in sub-Saharan Africa, but as very few ringing recoveries from that vast region are available precise areas remain largely unknown (Reichlin et al. 2009). Some populations fly only as far as the Mediterranean (van Wijk et al. 2013). Asiatic Russian and Central Asian populations move into milder regions to the south.

In southern Europe, *tschusii* (described by Kleinschmidt in 1907) breeds in Corsica, Sardinia, Sicily, mainland Italy and the western (Adriatic coast) Balkans. This subspecies is thought to winter in the southern Mediterranean and eastern Africa, for instance in Ethiopia and Eritrea.

In the very north of Africa (Algeria, north-west Tunisia), *mauretanica* (described by Rothschild in 1909) is resident. Some authors have suggested that it may also occur in Sicily, but this has not been verified (Brichetti and Fracasso 2020). In Asia, *himalayana* (Vaurie 1959) breeds in the north-west Himalayas, from northern Pakistan (perhaps eastern Afghanistan) to Kashmir and Himachal Pradesh, and is thought to winter to the south of the Himalayas in India and south-east Xizang, China, but again its precise distribution is unclear (Rasmussen and Anderton 2005). Other subspecies that have been described are: *sarudnyi* (described by Loudon in 1912), breeding in western Siberia from the Ural Mountains to the River Yenisey; *chinensis* (described by Hesse in 1911), in eastern Siberia and north-east and central China (MacKinnon and Phillipps 2000); and *japonica* (described by Bonaparte in 1850), in Japan, which breeds on the Kurile Islands, Hokkaido and northern Honshu, and winters from central Honshu to the Kyushu islands (Chikara 2019). The following are no longer recognised and have been assimilated into the above subspecies: *hyrcana* (described by Zarudny in 1913), *harterti* (described by Poljakov in 1915), *hokkaidi* (described by Kuroda in 1921), *pallidior* (described by Rensch in 1924), *intermedia* (described by Stegmann in 1927) and *incognita* (described by Stachanov in 1933).

It should be noted that the geographical distributional limits of most subspecies are vague, and few of their populations are geographically isolated. For instance, the borders of the ranges of *torquilla* and *tschusii* in the Balkans and Mediterranean are unclear. In the Alps of northern Italy, these two may well come into contact (Brichetti and Fracasso 2020). If so, they will most likely interbreed. Similarly, in Asia the range limits of *torquilla* and *himalayana* are ambiguous, and in Kazakhstan it is believed intergrades between *sarudnyi* and *torquilla* may occur (Wassink 2015). As yet, the question of intergrades between subspecies has not been adequately addressed. Furthermore, the fact that the majority of Wryneck populations are migratory means that the subspecies almost certainly coincide on passage and in wintering quarters. In addition, all those described are extremely similar in plumage and, to complicate matters even further, there are clinal as well as individual plumage variations within subspecies' populations. Hence the tendency to lump *sarudnyi*, *chinensis* and *japonica* with the nominate *torquilla*. All in all, the racial taxonomy of the Wryneck is probably in need of review. For descriptions of the recognised subspecies see Chapter 3.

Chapter 2

Anatomy and Morphology

As briefly mentioned in the previous chapter, woodpeckers have an anatomy that enables them to exploit ecological niches inaccessible to most other birds. Some exceptional studies on the anatomy of woodpeckers have been carried out, and these have highlighted how their bodily structures correlate superbly with their morphology (Goodge 1972; Bock 2015). These features relate principally to their foraging methods and nesting habits and typically include a strong chisel-shaped bill (suitable for boring into wood), strong, curved claws and stocky legs (needed for climbing up and clinging to vertical surfaces), rigid tail-feather shafts (which act as props that support them when they work on tree trunks), an extensible and retractable barbed tongue (to extract insect prey from deep within burrows) and a hyoid apparatus (bones

FIGURE 2.1 The long, extended tongue of a Wryneck (TH).

inside the tongue which allow its extension through and retraction from the bill when feeding). An anatomical adaptation that in the avian world is unique to woodpeckers is a frontal overhang of the braincase. This functions as a shock absorber, dissipating and diverting harmful vibrations from the impacts sustained when woodpeckers peck and drum on hard wood.

Of course, with somewhere between 220 and 250 picid species globally (depending upon which checklist is followed), there are variations in how these features evolved in different woodpeckers. The two wrynecks in the genus *Jynx* are the woodpeckers that break the mould, being anatomically and morphologically quite different from the tiny piculets and the generally larger true woodpeckers. Nonetheless, they do have some features in common with their relatives, such as their hyoid apparatus. This is similarly elongated, the bony horns extending up and over the skull. The hyoid, and associated long tongue, probably evolved first and were used to extract prey from crevices and holes, which indicates that excavation adaptations and habits developed later. This notwithstanding, the following are some of the anatomical features of wrynecks which differ from those of other woodpeckers.

Skeleton: Those woodpeckers that habitually peck into and drum upon trees have evolved structural modifications to their skeleton. For example, they have larger and broader ribs than most other birds. Robust ribs are most developed in those species which are highly arboreal and excavate their own nest cavities and least developed in those which do little excavation and do not forage by hacking into trees, such as Wrynecks. The anterior ribs of Wrynecks show no widening when compared with their pectoral ribs, whereas in piculets and true woodpeckers they are strikingly modified (Kirby 1980). Other skeletal features that are less developed in Wrynecks in comparison with most other woodpeckers include the shoulder girdle bones which are thinner, and the scapulae which are shaped more like those of the honeyguides (Höfling and Alvarenga 2001).

Skull: Wrynecks have a less reinforced skull than the true woodpeckers (Bock 2015). It is more delicate, lacking impact-buffering qualities.

Bill: The Wryneck's bill is relatively short, weak, lacks grooves and is slightly curved along the culmen. Compared with the bills of most other woodpeckers, it is narrow, slender at the base, pointed rather than chisel-tipped and therefore unsuitable for pecking into hard wood.

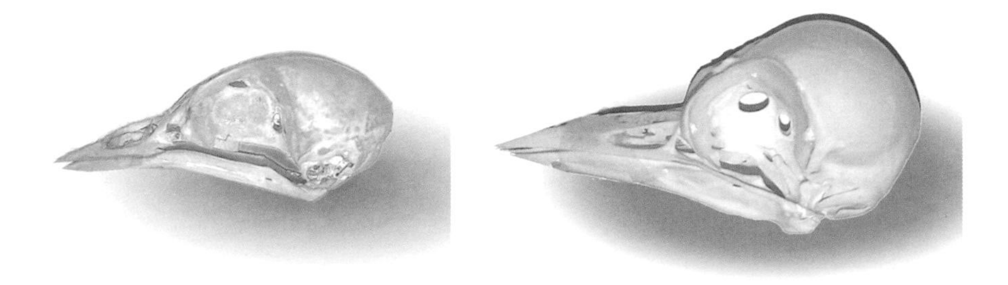

FIGURE 2.2 Comparison of the skulls and bills of Wryneck, left, and Great Spotted Woodpecker, right (TP).

FIGURE 2.3 The Wryneck's bill is slender and pointed rather than chisel-tipped like those of its true woodpecker relatives (TH).

FIGURE 2.4 Left foot and claws. Like most woodpeckers, Wrynecks have four toes on each foot (TH).

Legs and feet: Wrynecks have four toes on each foot, with a zygodactylous arrangement (first and fourth toes directed backward, second and third forward). Although all woodpeckers have their feet in this yoke-shape, it is in fact more suited for perching on branches than for climbing tree trunks, and therefore ideal for Wrynecks. Structurally, the limbs are one of the few anatomical features these birds possess that are typical of the woodpecker family in general. Nevertheless, the claws are less curved than those of most woodpeckers, which illustrates how the species is less adapted to climbing than its relatives.

Muscles: On the whole, Wrynecks are not as physically robust as typical woodpeckers. This is reflected in their muscles, which are not so dense or tough as those of their relatives (Kirby 1980).

Glands: The nasal glands of woodpeckers have evolved to facilitate working on wood. Wrynecks differ from the true woodpeckers in the form, size and position of these glands. For instance, their nasal gland is located in the orbit and their maxillary gland, which is large in the majority of woodpeckers, is small (Goodge 1972).

Nostrils and nasal bristles: The nostrils are more oval and open than those of other woodpeckers and only partly covered with feathers or bristles. As they do not excavate wood, Wrynecks have no need for such protection from wood particles.

FIGURE 2.5 Comparison of the tails of Wryneck and Great Spotted Woodpecker *Dendrocopos major*. Specimens in the Linz Biologiezentrum, Austria (SW).

Tail: Wrynecks have a relatively long tail with soft feathers which are rounded at their tips. The feather shafts (rachises) are also less rigid than those of true woodpeckers. Consequently, the tail does not function as a brace as it does for those species that cling to trees (although they will splay the tail on the surface below the nest hole to provide some support when feeding chicks at the cavity entrance).

Tongue: The tip of a woodpecker tongue varies in structure among species. Some are pointed, for impaling invertebrates, and others are rounded, for licking up prey. To varying extents, most are lined with barbs. Though Wrynecks lack many of the adaptations their relatives possess, they do have a long tongue which, together with it being only slightly pointed, smooth and lacking barbs, relates to the foraging techniques used and kinds of prey sought. In a classic work, Leiber (1907) states that Wryneck and Great Spotted Woodpecker have tongues of about equal length, and that the Great Spotted's is some 2.5 times as long as its bill. From that it can be derived that the total length of the Wryneck's is approximately 70 mm.

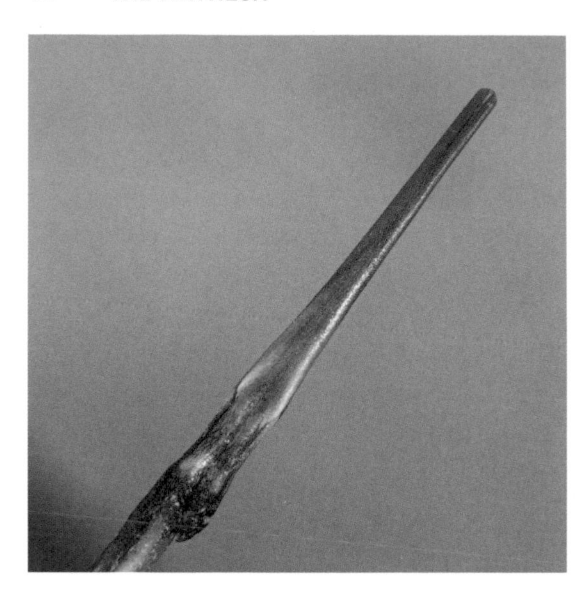

FIGURE 2.6 Tip of a Wryneck tongue from above. Smooth and lacking barbs. Specimen in the Linz Biologiezentrum, Austria (SW).

FIGURE 2.7 A Wryneck 'yawns' and reveals a little of its long and remarkable tongue. May 2021, Kocsér, Hungary (RP).

Chapter 3

Description and Identification

Wrynecks are often referred to as atypical woodpeckers. They are regarded as unusual because, in terms of their structure – being slim with a relatively small head, a fine bill and a long tail that is fan-shaped and more or less rounded at the tip – they do not resemble the typical woodpeckers that everyone recognises, and which make up the subfamily Picinae (see Chapter 1).

FIGURE 3.1 Perched Wrynecks often have a songbird-like jizz. April 2020, Jihlava, Czech Republic (AT).

Jizz

These birds are often seen on the ground where they forage inconspicuously on short-grass and stony areas, among tussocks and at terrestrial ant nests. They typically hop in jerky movements, often with their tail slightly cocked. When in trees, they are usually more sluggish in their movements and they tend to perch crosswise on branches in the manner of songbirds. Unlike most woodpeckers, they seldom shuffle along boughs or cling to the sides of tree trunks, and only sporadically use their tail to support themselves. They usually fly low to the ground, the flight pattern being fairly direct and straight, although they may swoop and briefly close their wings. With their long tail obvious, they can appear more like a small thrush, shrike or large warbler. Upon landing, they occasionally flick their wings briefly. When alert they will sometimes adopt a stretched-out posture, and when alarmed they raise their crown feathers into a 'punk' crest. When observed in these postures Wrynecks are often described as being 'reptilian' (Gorman 2004).

Similar species

In most situations Wrynecks are unmistakable. Realistically, they cannot be mistaken for any other woodpecker except their close relative the Red-throated Wryneck, with which they co-occur in parts of Africa in winter. In breeding areas in Europe and Asia, they are more likely to be confused with a songbird. In some situations, such as when a Wryneck darts from one bush to another, it might be mistaken for a female or juvenile Red-backed Shrike *Lanius collurio* or Barred Warbler *Sylvia nisoria* owing to their similar size and proportions. The plumage patterns of both wryneck species, with cryptic browns, greys, black and white in the colouration, are often compared to those of nightjars (Caprimulgidae) and the plumage has even been described as 'caprimulgiformlike' (Short 1982), but wrynecks cannot really be confused with that family owing to the relative sizes, structures and entirely different behaviours of the two.

Measurements

The following biometric figures are for adult Wrynecks. They are based on Cramp (1985), Glutz and Bauer (1994), Baker (2016), Demongin (2016), and measurements taken by the author in museum collections in Europe. Measurements from male and female Wrynecks are combined, as sexual

FIGURE 3.2 Body lengths vary greatly as can be seen from these two adult specimens in the Budapest Natural History Museum, Hungary (GG). Both of these birds were collected at the same site (Ohat, Hungary) in spring, the shorter bird on 26 April 1959 and the longer on 1 May 1957. The longer is 195 mm long, from bill-tip to tail-tip, and the shorter 150 mm – a considerable difference of 45 mm.

differences are slight and the sex of specimens was not always accurately documented. Figures for subspecies also are combined, as samples were small and, in some cases, the subspecific identification was uncertain.

- Overall length from bill-tip to tail-tip: 150–195 mm. There can be significant differences in the lengths of individuals, with the tail around a third of overall length.
- Tail length: rather variable, 50–70 mm.
- Wing length: rather variable, 75–94 mm.
- Wingspan: rather variable, 250–270 mm.
- Bill length, from forehead to tip: variable, 14–17.5 mm.
- Tarsus: fairly constant, 18–21 mm.

Plumage

Though subtly well marked, both sexes can, from brief or distant views, appear uniformly brownish-greyish with few obviously distinguishing features. Wrynecks are cryptically patterned with an intricate mosaic of brown, fawn, grey, black, yellow and buff streaking, barring and mottling, from the crown to the rump and tail. This is sometimes described as a 'tree-bark pattern' (although this is misleading as wrynecks of both species spend much of their

FIGURE 3.3 Adult Wryneck. April 2010, Norfolk, England (NeB).

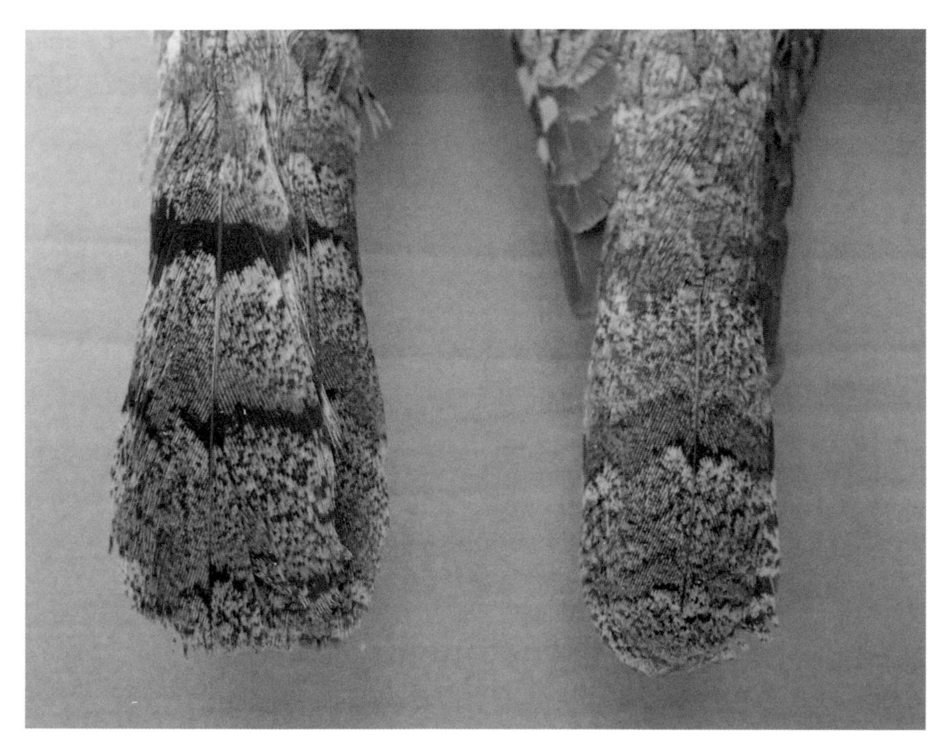

FIGURE 3.4 These two examples of adult Wrynecks, in the collection of the Budapest Natural History Museum, Hungary, illustrate the variation that can occur in uppertail barring (GG). One has two distinct, dark bars, one has two undefined bars.

time on the ground rather than against tree bark). Many feathers have pale tips, but these are not always noticeable in the field. A blackish-brown band runs from the crown over the nape and mantle, between the scapulars, and down the back. This band often widens into a diamond-shape on the mantle. The scapulars have pale edges and dark centres. The sides of the neck are grey as are two 'braces' on the mantle. The lower back and rump are lightly barred with brown. The tail has several (often four) widely spaced dark bands, although some may be faint or broken and the innermost band may be hidden by the uppertail-coverts. The undertail-coverts are creamy white and lightly marked with black flecks.

The crown is finely barred with black and edged with dark brown. A broad brown stripe runs over the eye, across the ear-coverts and down on to the side of the neck. Some individuals show a faint, pale malar (moustachial) stripe, which is thinly barred with black. The cheeks, throat, neck and chest are yellowish-sandy, finely barred with brown. The breast and flanks are cream, marked with dark flecks and arrowhead and chevron shapes. The more vivid colour of the throat and chest may merge with the paler breast, or the two areas can be well demarcated. The belly and ventral area are cream or white and much less marked, often plain. The wings are mainly brown with the darker flight feathers marked with rufous and buff spots.

Bare parts

The bill is horn-grey, sometimes brownish, relatively short with a fine, pointed tip and a curved culmen, and is narrow across the nostrils. Unlike the bills of other woodpeckers, it is smooth and lacks ridges and grooves. The tongue is pink. The irides of adults are chestnut in colour, those of juveniles greyer and those of nestlings all dark (pers. obs.). Second-year individuals usually have transitionally coloured irides, reddish but greyer/browner at the outer edges (Baker 2016). Wrynecks have short legs and their four-toed feet are greyish-brown, sometimes with a greenish tinge.

The sexes

Wrynecks are not sexually dimorphic. Indeed, males and females appear virtually identical (monomorphic) and are usually inseparable in the field.

Males: In the breeding season, adult males are generally a brighter, richer yellow, even ochre, on the throat and chest, compared with females. This is, however, rather variable and noticeable only in some pairs and is therefore ultimately unreliable as a diagnostic sexual-identification feature (Gorman 2004).

FIGURE 3.5 Adults have chestnut-coloured irides. September 2014, Spurn, Yorkshire, England (CG).

FIGURE 3.6 First-winter and second-calendar birds have grey irides. September 2018, Bourgas, Bulgaria (GG).

Females: In addition to having the yellow areas of their plumage sometimes less vivid, females are also slightly smaller than males. Perhaps unsurprisingly, this, too, is noticeable only when the two members of a breeding pair are seen together and, in many cases, is not obvious at all.

Juveniles and immature birds

Young Wrynecks look much like adults although they tend to have more white barring on the crown, brown areas are duller, the throat is greyer, the rump is cream-coloured and the tertials are rustier. There is also a white spot near the tip of each tertial (in adults these are cinnamon in colour) but these are difficult to observe in the field. They are also much less streaked and barred below than adults, and the markings on their throat, chest and flanks are paler,

FIGURE 3.7 A juvenile a few weeks after fledgling. July 2015, Pescasseroli, Italy (PH).

greyish rather than blackish. They typically lack chevrons on the underparts and have fewer but bolder black bars on the tail. Structurally, they also appear more compact than adults as they have a shorter tail and more rounded wings, and their outer primary is longer (Gorman 2004).

Subspecies identification

Separating the various subspecies in the field on the basis of plumage is all but impossible, despite the literature often detailing differences. In addition to all subspecies being essentially similar in appearance, there are also clinal and individual variations in size to consider. One study found that the wings of birds in central Europe tend to be longer and more pointed and tails shorter than birds in south-west Europe, although all were the nominate *torquilla*; the same trend was noted for birds in East Asia compared with those in Central Asia (Eck and Geidel 1974). In practice, subspecific differences are likely to be

FIGURE 3.8 Adult *tschusii*. June 2020, Modena, Italy (ET). Truth be told, the various subspecies are exceedingly difficult to separate in the field.

detectable only by examining a series of museum skins of birds of the same age and season. Nevertheless, the following distinctions are usually described.

Subspecies *tschusii* is similar to the nominate but darker overall, particularly on the tail and the stripe down the upperparts; this stripe usually runs from the central crown to the mantle on *torquilla*, but on *tschusii* it may start from the forecrown and is often bolder. On *tschusii*, the throat, chest and upper breast are more heavily barred with black-brown and the black flecks on the mantle, scapulars and crown are also bolder and broader; there is more barring on the undertail-coverts and vent and the flanks have fewer chevrons, being barred rather than scaled. The wings of *tschusii* are on average slightly shorter and more rounded. Form *mauretanica* is said to be smaller than the nominate, darker above, creamier on the throat and chest, and less marked overall; *himalayana* is apparently more heavily and extensively barred below; *sarudnyi* is described as plainer and greyer above, whiter and less marked below, and paler on the throat; *chinensis* is said to be darker, heavily barred and smaller than European birds; *japonica* is claimed to be somewhat rufous and more barred than *torquilla*, but it should be noted that it is not unusual to find reddish individuals in Europe (Winkler et al. 1995; Gorman 2014).

Hybrids and aberrants

There seems not to be any verified record of Wrynecks interbreeding with other bird species. A possible *J. torquilla* × *ruficollis* hybrid reported from Cameroon (Desfayes 1969) was later judged to have been an aberrant Red-throated Wryneck (Short and Bock 1972).

Chapter 4

Moult, Ageing and Sexing

The majority of birds moult all their feathers at least once a year, usually after they have finished breeding: this is termed a *post-breeding moult*. Most woodpeckers, on the other hand, do not renew all their feathers each year. Although they replace their primaries, woodpeckers retain some of their primary coverts and some secondaries during their post-breeding moult, so the renewal of the primary coverts is not linked to the corresponding primary, as it is in many species (Ginn and Melville 1983). This is significant, because the retention of these feathers can help in the ageing of individuals and consequently reveal information on their demographics (Laesser and van Wijk 2017). Ultimately, as with other birds, the moult pattern of woodpeckers is correlated with their foraging habits, breeding cycle and, in the case of the Wryneck, also with migration.

Tail feathers

As most woodpeckers depend upon their tail feathers (rectrices), particularly the robust central pair, to support them on tree trunks when excavating and foraging, they have evolved a distinctive tail-moult strategy. Rather than replacing their central feathers (R1) first as most birds do, woodpeckers first moult the second pair (R2), leaving the central pair, which is so important to them, until last. Wrynecks do not moult their tails in this way. Rather, they replace their rectrices in reverse sequence (ascendant moult), replacing the outermost two (R6) first (Stresemann and Stresemann 1966). Still, those of adults quickly wear in the breeding season and this may be due to abrasion from foraging on the ground and when the tail is pressed against the surface below the nest hole entrance when feeding chicks. Rectrices can differ significantly in length between individual Wrynecks (Hansen and Synnatzschke

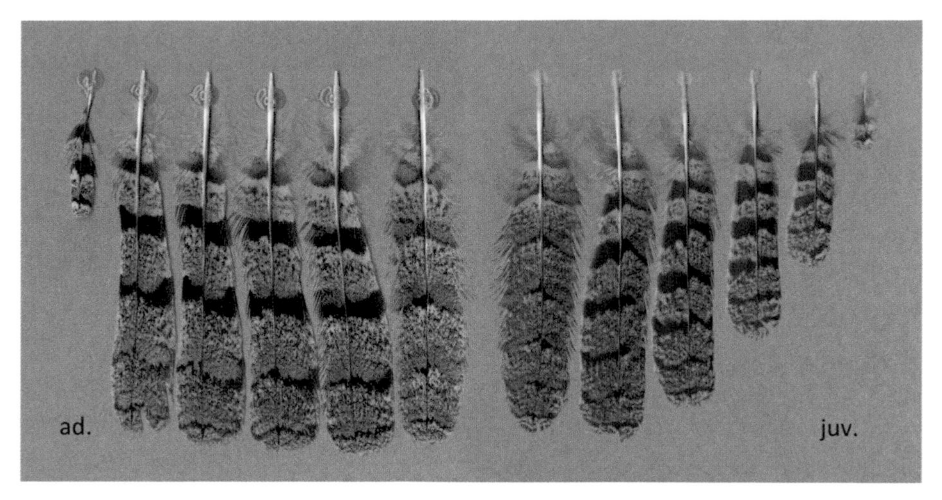

FIGURE 4.1 Comparison of juvenile and post-juvenile (adult) rectrices. Juvenile birds show a stepped tail, whereas in older birds only the outermost feather (R6) is noticeably smaller than the others. Longest juvenile rectrix (R2) 71 mm; adult (R2) 73 mm. The juvenile was found dead, of unknown cause, in July 1988 at Eschenau; the adult was killed by a domestic cat in May 1991 in Traisen, both in Lower Austria (TH).

2015). Tail patterns, too, especially the number, extent and shape of the dark brown-black bars on feathers R2 to R5, and the presence of smaller black or white bars, vary greatly but are not related to sex or age. It has been found that tail patterns are consistent over the years and reflect those of a bird's parents (Becker et al. 2013).

Moult patterns

It is clear that there are gaps in our knowledge of Wryneck moulting patterns. What follows is an attempt to present the current information that is available on this challenging subject. Wrynecks have 10 primaries per wing (the outermost is minute after the post-juvenile moult), 11 secondaries per wing and 12 rectrices. The species undergoes two moults per year, an almost complete one in summer (post-breeding) and a partial one in winter (pre-breeding), changing the body feathers twice (Jenni and Winkler 2020). The overall moult pattern is more complex than that of other picids, and errors in ageing birds are easily made. Some matters have, however, been clarified recently through a four-year analysis of the moult patterns of breeding Wrynecks in Switzerland (Laesser and van Wijk 2017). These researchers established that the information on moult and ageing in much of the existing literature is occasionally ambiguous. For example, adult Wrynecks can divide the moult of their secondaries between their post-breeding and pre-breeding phases, and so they sometimes exhibit two generations of feathers. Indeed, even people experienced in handling the species are often averse to precisely ageing Wrynecks unless the individual birds concerned are known to have been ringed as nestlings or are second-year birds, and these two stages are considered to be the only times when age can be determined with accuracy.

As already mentioned, the fact that Wrynecks are migratory means that an understanding of their moult strategy can help in an ecological context. It has been shown with birds in general that, when they retain their primary coverts, they moult their primaries at a faster rate (Kiat and Izhaki 2017). Such a strategy perhaps benefits northern populations of Wrynecks, which have large distances to travel before and after breeding, nest relatively late in the year, and therefore have limited time to moult before local conditions deteriorate (Iovchenko and Kovalev 2005). Of course, some populations have more time than others, as some are short-distance migrants and others are not migratory at all (see Chapter 10). Migratory individuals start moulting in their breeding areas once nesting is over, some continuing to moult in

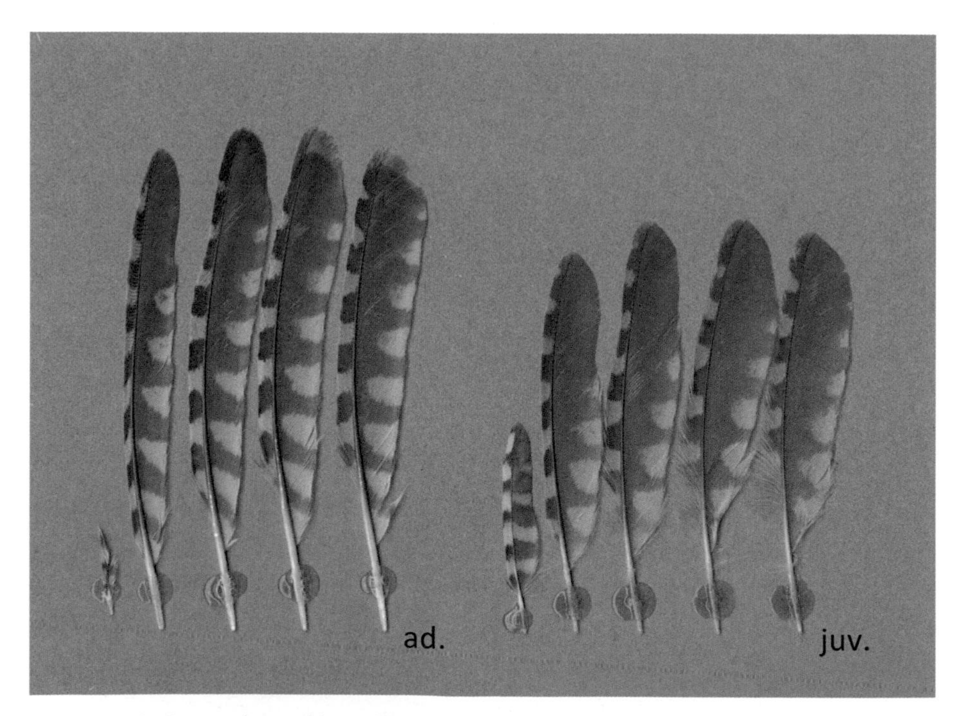

FIGURE 4.2 Comparison of juvenile and post-juvenile (adult) primaries. The outermost primary (P10) in the juvenile is about half the length of P9, whereas in post-juvenile plumages this feather is much reduced. Juvenile primaries are clearly shorter than in subsequent plumages. Longest juvenile primary (P7) 61 mm; adult (P7) 74 mm (TH).

staging areas while on migration. Others complete the moult in their wintering quarters after having suspended it during migration. Some individuals may use all of these strategies. All things considered, more study on moult in different populations (especially non-migratory ones) is needed before any firm conclusions can be drawn. Laesser and van Wijk (2017) found that they typically delayed replacing their primary coverts, with none renewed during their post-juvenile moult. At most just one or two (exceptionally three) distal primary coverts were replaced by the first complete moult at around one year of age, and then again, a limited number were replaced in the second post-breeding moult, which Wrynecks in their third calendar year undergo. Three to six primary coverts were found to have been moulted at more advanced ages. The fact that some birds examined in this study seemed not to replace their first (juvenile) primary coverts until their sixth calendar year is remarkable, especially as Wrynecks have a relatively short lifespan. This postponed moult results, to a certain extent, in the primary coverts being useful for estimating different ages.

First calendar year (fledgling/juvenile plumage): Juveniles, before their first wing moult, have the rather long outermost primary (P10) half the length of the ninth (P9), whereas on adults it is tiny. The juvenile P10 is much longer than the one that replaces it, while in other woodpeckers this size difference is slight. The longest primaries in the juvenile stage are about 10 mm shorter than in subsequent plumages (Sutter 1941; Jenni and Winkler 2020). Juvenile secondaries are, on the other hand, 1–2 mm longer than those of adults, and juvenile tertials have two white spots near their tip, although often only the spot on the outer web is white with the spot on the inner web being brownish (Winkler 2013). The primary coverts are narrow and pointed, often with some faint cinnamon spots, and they may have arrowhead-shaped markings on their tips (Baker 2016). The tail of juveniles is stepped, with the distal rectrices (R3–6) gradually reduced in length (Hansen and Synnatzschke 2015). On adults, R1–5 are nearly the same in size.

Post-juvenile moult: Juveniles usually begin their first moult, starting with the innermost primary (P1), while still in the nest cavity, shortly before fledging (Demongin 2016). In central Europe this can be from mid-June to mid-July depending upon location and whether the birds are from a first or a second brood. It is completed after fledging, usually by late July or August but sometimes in September, again depending upon such factors as locality. In northern latitudes, where Wrynecks are a relatively late-breeding species, newly fledged young have less time to prepare for migration, before autumn approaches, than do those breeding farther south. For instance, a study conducted at the Lake Ladoga Ornithological Station in north-west Russia, using data from 1968 until 2002, found that juveniles started their moult at between 19 and 23 days of age, that is, when still in the nest or a few days after fledging. The earliest recorded date when this first moult commenced was 4 July and the latest was 18 July. The duration was 60 to 70 days (Iovchenko and Kovalev 2005), which is rather short when compared with the moult duration of other woodpecker species. The partial post-juvenile moult involves the primary, tail and body feathers. It seems that juveniles do not moult their secondaries and tertials, nor their primary coverts in one go (Miettinen 2002; Iovchenko and Kovalev 2005; Laesser and van Wijk 2017). Some authors state that the central tail feathers, the second pair and sometimes even the third pair can be retained (Sutter 1941; Winkler 2013). Most individuals in migratory populations continue to moult as they begin to move southwards in August and September. At Lake Ladoga, birds of unknown origin which are trapped, including many juveniles which are believed not to be local birds, stop and moult there from late July into August before flying on (Iovchenko and Kovalev 2016).

First calendar year (first-winter plumage): After the post-juvenile moult Wrynecks can be aged by observing that all the primary coverts are of the same generation, whereas in all other age categories except first-summer plumage, there is invariably a mix of old and new primary coverts. The tertials often still show a white spot near their tips (Winkler 2013; Laesser and van Wijk 2017) and the irides are greyish-brown (Blasco-Zumeta and Heinze 2017). Young birds resume this moult in their wintering areas, where they replace most, sometimes all, of the secondaries. In exceptional cases some juveniles retain all of their secondaries (Winkler 2013). Furthermore, a pre-breeding moult is undertaken in early spring (before migration), involving the wing-coverts (but not primary coverts) and most or all of the tail and body feathers (Stresemann and Stresemann 1966; Demongin 2016). In general appearance, and owing to their new secondaries, young Wrynecks resemble older ones, but they can be correctly aged once they return to breeding sites in the following spring.

Second calendar year (first-summer plumage): In their first summer, most Wrynecks have retained some secondaries and tertials from their juvenile plumage. When some juvenile secondaries are present ageing is relatively straightforward. Juvenile secondaries are more worn and longer, so they protrude slightly beyond those newly acquired. Second-year birds have only juvenile primary coverts, which are pointed at the tip, similar in hue to the

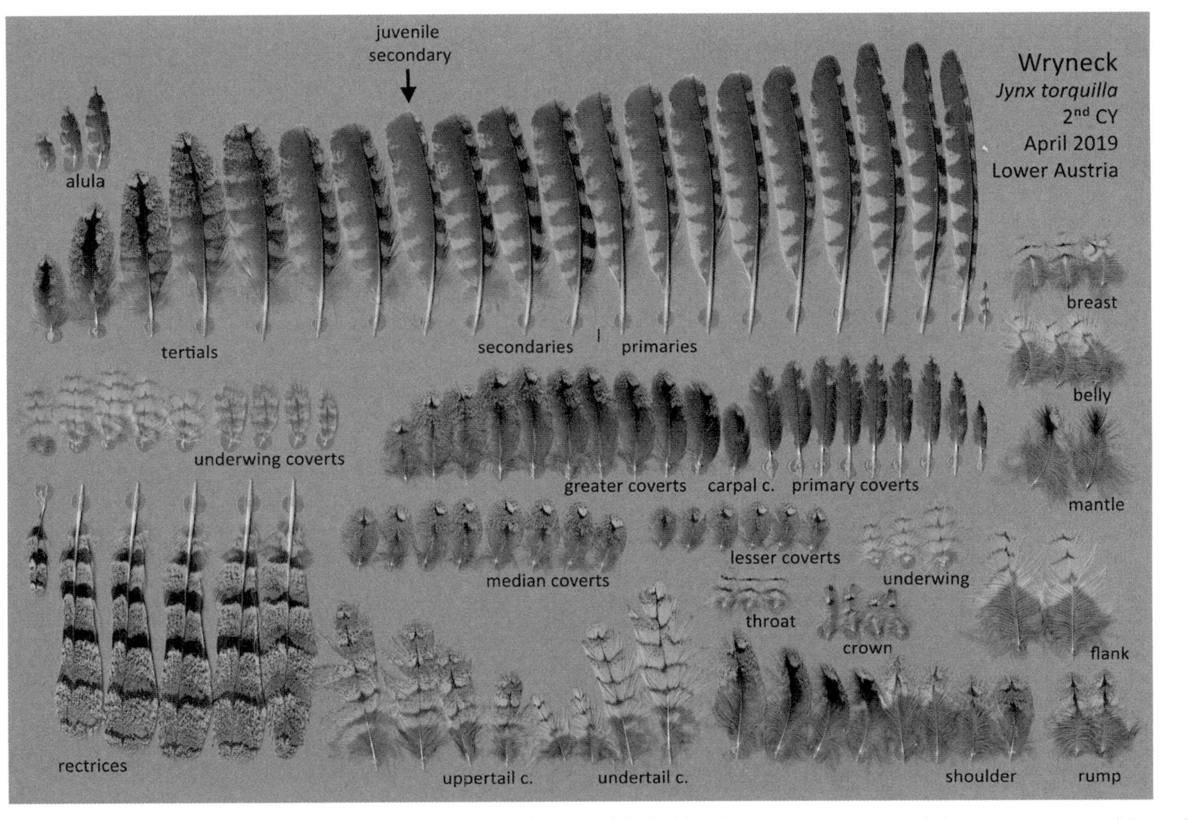

FIGURE 4.3 Feather-plate of a second calendar year, first-summer plumage bird. All primary coverts are of the same age and have been retained from juvenile plumage (lacking white spots). The slightly longer and narrower fourth secondary (S4) has also been retained, which makes ageing of this bird fairly straightforward. Longest primary (P7) 74.5 mm; longest rectrix (R2) 72.5 mm. The bird was found dead of unknown cause, in April 2019 in Haringsee, Lower Austria (TH).

FIGURE 4.4 This bird is probably in the latter stages of its second calendar year, after or during its first post-breeding moult, as its PC8 is new. PC9 is not clearly visible in this image. Also note the greyish iris. September 2018, Bourgas, Bulgaria (GG).

primaries, and, although they may be moderately abraded, they often look quite fresh. The lack of any new primary coverts (all nine being of the same generation) is a key plumage feature in separating second-year individuals from older ones (Laesser and van Wijk 2017). The small distal primary covert (PC 9), however, should be carefully inspected (as it is easily overlooked) when no juvenile secondaries are left to rule out third-year birds. The iris colour of second-year individuals is often lighter, too, with outer areas greyish (Blasco-Zumeta and Heinze 2017).

Second calendar year (first post-breeding moult): Soon after breeding, these non-juvenile birds undertake an almost complete moult which, in Europe, is usually completed by September. The renewal of the primaries starts from the innermost (P1) and proceeds in a descending direction, while divergent renewal of the secondaries begins at S8 and then another moult pattern begins ascending from S1. The rectrices are also replaced ascendantly. There are exceptions, as some feathers (secondaries and the majority of primary coverts) may be retained, the coverts being important for determining age. The majority of Wrynecks renew only one or two of their outermost primary coverts (PC9 and PC8), although some may replace one or two more and occasionally all may be retained. These feathers are more or less pointed, the pale spot near their tips usually being still obvious, and in hue they contrast with the primaries (Laesser and van Wijk 2017). Secondaries that have been retained in summer can be renewed during the winter partial moult, but it is also not unusual to find two generations of secondaries on older birds (as with other woodpeckers), although older secondaries do not protrude as they do in the first-summer plumage.

Third calendar year (second post-breeding moult): Even individuals that have completed a second nesting season usually retain a number of their juvenile primary coverts. Indeed, the number of primary coverts renewed during the second post-breeding moult is again often limited to just the outermost ones. Retained juvenile primary coverts can be distinguished from those of second-year birds as, being by now three years old, they are significantly faded and have heavily abraded pale spots. Nevertheless, the ageing of third-year birds on the basis of the moult of this feather tract is problematic and requires experience. After more than two years of wear the primary coverts are typically more faded than the primaries. The iris colour at this age is typically reddish-brown-chestnut (Demongin 2016).

Older birds: Some juvenile primary coverts are retained until at least the third breeding season (fourth calendar year) and in some cases even longer, hence accurate ageing continues to be challenging. Those retained are extremely abraded and more faded than the primaries and can be distinguished from worn adult primary coverts owing to their brownish tinge and uneven fringes resulting from heavier wear. Four- and five-year-old individuals typically renew more primary coverts than second- and third-year birds do during their post-breeding moults, but finding individuals that have moulted all of these feathers at the same time is rare. As the complete moult of the primary coverts spans many years, it is commonplace for Wrynecks as old as four, five and perhaps even older to have juvenile feathers in their plumage (Laesser and Wijk 2017). According to the same authors, individuals showing two adult generations of primary coverts (three generations in total) can be aged as being three years old at least. Those birds that no longer have any juvenile feathers can be aged with a higher degree of certainty as being at least four years old.

Despite much of the literature stating that a single spot on the primary coverts is diagnostic of adults (van Duivendijk 2011; Baker 2016; Demongin 2016), it is not certain that this is the case. Although some juveniles may have arrowheads on these coverts, others can have an adult-like pattern (Laesser and van Wijk 2017). In short, the ageing of Wrynecks in the hand after their post-juvenile moult is probably best attempted by looking at the condition of the primary coverts to see whether they are faded or worn, whether any have been renewed, and whether there are contrasts between them and the primaries.

Finally, it should be remembered that Laesser and van Wijk's (2017) study that focused on primary coverts resulted from work with a migrating population of Wrynecks in Switzerland. It is not yet clear if more southerly or resident populations moult in the same way.

Nestlings

Knowing the ages of nestlings is valuable for studies on breeding biology and phenology, but accurately determining the ages of chicks in the nest is often difficult. Researchers who examined Wryneck broods in nest boxes found that chicks can be aged through careful and regular scrutiny of their growing plumage and by taking wing and weight measurements (Mulhauser and Zimmermann 2014). The same authors found they could obtain accurate estimates for chicks aged between 2 and 16 days, but after that, around the time they fledge at 17 to 22 days, ageing became less reliable. They noted the following development of features. Newly hatched birds were featherless and the first feather growth, in the form of a line of black dots, began to show on

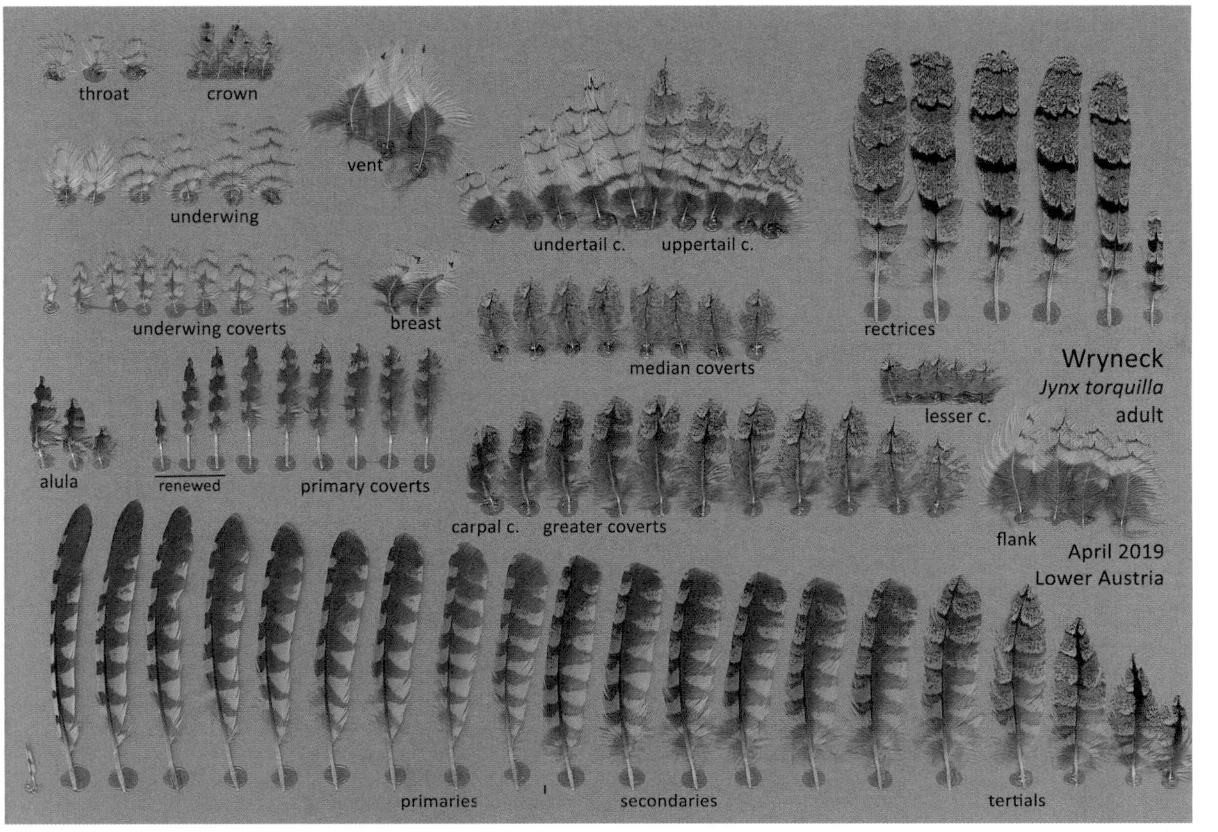

FIGURE 4.5 Feather-plate of an adult. Only the outermost three primary coverts have been renewed. The inner PCs are probably of the same generation. This bird can be aged as being (at least) in its third calendar year. Longest primary (P8) 74.5 mm; longest rectrices (R1–R3) 69 mm. Found dead after colliding with a wind turbine in April 2019, in Prottes, Lower Austria (TH).

the forewing after just 2 days. By day 3 black dots (pins) had appeared on the remiges, rectrices, and greater and median coverts. By day 4, V-shaped marks began to appear on the nestlings' backs, which had changed to Y-shapes by day 6. At about this time the eyes began to open, and the chicks developed obvious dark ocular masks of skin. Around days 7 and 8, the first flight and tail feathers were emerging. By day 11, brownish primary and secondary remiges were mostly out of their sheaths and the back was speckled with beige. By day 12, the triangular-shaped patch of bare skin in the centre of the back was hardly visible. On day 13, the sheaths of the flight feathers and rectrices were still clearly visible. By day 15, brown marks on the head had developed into lines. By day 16, a few individuals still showed rectrices with sheaths visible at their bases.

Similar work by Sutter (1941) mentions black dots on the body feathers on day 4 and that by day 13 the whole of the body is feathered.

FIGURE 4.6 A 9- or 10-day-old nestling about to be ringed. June 2006, Milovice, South Moravia, Czech Republic (JC).

FIGURE 4.7 A 14- or 15-day-old nestling. The already well-developed feet and claws are typical of woodpeckers. June 2021, Balatonmáriafürdő, Hungary (GG).

Longevity

The oldest known Wrynecks have been documented by ringing recoveries. The two oldest recorded are an individual aged six years and ten months, controlled by a ringer in the Czech Republic, and one that was shot in Finland at the age of six years two months (Fransson et al. 2017). Of course, some individuals no doubt live for longer than the two mentioned here, perhaps up to 10 years (Winkler et al. 2020), but this has not been documented. Curiously, it is likely therefore that many Wrynecks perish before ever replacing all of their juvenile plumage.

Sexing

Determining the sex of Wrynecks in the hand (as in the field) is difficult. Indeed, most ringers do not even attempt it as there are, for example, overlaps in biometrics (Yosef and Zduniak 2011). Some who do try to sex them have tended to use two criteria: the presence of a brood patch and/or cloacal protuberance. The first is a patch of swollen, reddish skin on the belly that develops just prior to the eggs being laid. Once the chicks have hatched it begins to recede and is finally feathered again during the next moult. The second is a bulbous swelling of the cloaca which occurs in male birds (Baker

2016). But there are problems with both of these features as indicators of the sex of Wrynecks. First, as both males and females incubate, both have a distinct brood patch (Menzel 1968) and finding one on a Wryneck is therefore meaningless when it comes to sexing. Second, the protuberance of a male can easily be confused with the distended cloaca of a female (Demongin 2016) and, perhaps more significantly, use of cloacal morphology has not yet been proven to be a reliable means of sexing woodpeckers of any kind. Finally, even if employment of this last method were appropriate, it would be useful only during the short breeding season when birds are sexually active.

Chapter 5

The Red-throated Wryneck

The genus *Jynx* contains just two species: the one that is the main subject of this book and the African endemic Red-throated Wryneck, the latter sometimes called the Red-breasted, Rufous-necked, Rufous-breasted, Rufous-throated or African Wryneck. These two relatives coincide in a few areas in sub-Saharan Africa when the Eurasian-breeding species migrates south to winter in, for instance, Cameroon, Nigeria, Uganda or Ethiopia's Rift Valley (Short 1988b). The breeding ranges of the two species do not coincide. Red-throated is not so widely studied as its Palearctic congener, but the two species seem to show little divergence in terms of their natural history. Ways in which Red-throated differs from Eurasian (other than in plumage and biometrics) include the fact that it is non-migratory, that it lays smaller clutches, and that its chicks have a shorter fledging period (Tarboton 1976).

Taxonomy

The Red-throated Wryneck was first described by Johann Georg Wagler in 1830 from the type-locality Uitenhage, Korrumus Mountains, Eastern Cape Province, South Africa (Short 1988b). Three subspecies are generally recognised (Short 1982). These are the nominate *ruficollis* (south-east Gabon, Congo to south-west and east Uganda, south-west Kenya, the very north of Tanzania, north Angola, north-west Zambia, north Mozambique, Eswatini (formerly known as Swaziland), eastern South Africa); *aequatorialis* (Ethiopian highlands); and *pulchricollis* (south-east Nigeria, Central African Republic, Cameroon to South Sudan, north-west Democratic Republic of the Congo and north-west Uganda). Other subspecies previously claimed but today not normally accepted include *pectoralis* from northern Angola, *diloloensis* from eastern Angola, *cosensi* from Kenya and Tanzania, and *thorbeckei* in Cameroon (del Hoyo and Collar 2014). For subspecies plumage descriptions see the section on Subspecies.

Identification

The two species are similar, particularly in shape and structure, and in brief views could be confused. Although they overlap in size, Red-throated Wrynecks are on average slightly larger than Eurasian Wrynecks, at up to 190 mm in length. They are browner overall, streaked rather than barred on the breast, belly and flanks, do not have a dark streak through the eye, are rufous on the ventral area and, most notably, have a reddish throat and breast. All things considered, Red-throated Wrynecks are unlikely to be confused with any other birds in their range.

The sexes

Male and female Red-throated Wrynecks are similar and difficult to separate in the field. Measurements taken of birds in the hand, however, have shown that males are, on average, slightly larger (2–3%), but with a shorter tail, and generally heavier (up to 10%) than females (Tarboton 1976).

General description

On both sexes the chin, throat and upper breast are rufous, varying in tone from red to rusty or chestnut, and the colours varying also in extent. The lores and ear-coverts are dusky, barred buff and brown, and the cheek and malar area are pale buff-white and vermiculated black. The lower breast and belly are creamy white and streaked chocolate brown. The flanks, undertail and ventral region are washed cinnamon, although once again this is variable in hue and extent. The underwing is buff with rufous-brown barring. The primaries are dark brown with boldly rufous bars, and the scapulars, tertials and wing-coverts are greyish-brown with finer barring and dark spots. The wing-coverts are white tipped. Above, from the crown to the uppertail-coverts, Red-throated Wrynecks are brown and subtly and finely mottled, speckled and flecked brown and black. Variable brown-black spotting which runs from the crown down on to the mantle is sometimes dense enough to form a streak. The rump and uppertail-coverts have fine black barring, the long grey tail being more boldly barred with black. The bill is fine and grey and the irides of adults are reddish-brown (chestnut), while the legs are grey, sometimes with an olive tinge.

FIGURE 5.1 Adult Red-throated Wryneck, subspecies *aequatorialis*. November 2019, Langano, Ethiopia (NiB).

FIGURE 5.2 A pair of Red-throated Wrynecks. November 2009, Lake Naivasha, Kenya (NeB).

Juveniles and immature birds

Young Red-throated Wrynecks look much like adults from as young as 20 days (Tarboton 1976). They are, however, darker and more heavily barred above than adults, lightly and finely barred below, and the rufous throat-and-chest patch is smaller, duller and vermiculated (Gorman 2014).

Subspecies

The subspecies differ mainly in the amount and tone of the rufous areas of plumage and, to a lesser extent, in their upperpart colouration. The red on the nominate *ruficollis* begins on the chin and extends down to the chest, where it usually ends abruptly. This red area is most extensive on *aequatorialis*, sometimes reaching from the bill down to the lower breast, and the flanks are washed cinnamon from the undertail-coverts up onto the lower belly. The subspecies *pulchricollis* has the smallest area of red, the chin lacking it and instead being white with fine brown barring, while red is confined to the lowermost throat and upper breast; in addition, the red is darker and more chestnut in tone compared with other subspecies. The upperparts also tend to be more rufous. It is important to note that there can be considerable individual variation in all regions, as well as clinal increases in size, usually in a northerly direction (Short 1982). Such differences have sometimes resulted in further subspecies being described and claimed on the basis of their evidently distinctive plumage features. For instance, '*thorbeckei*' in Cameroon was said to be more extensively barred on the throat than other races, but individuals in other populations can also show this feature. Other proposed subspecies '*pectoralis*', '*diloloensis*', '*striaticula*', '*rougeoti*' and '*cosensi*' look very much like the nominate *ruficollis* (Short 1982; del Hoyo and Collar 2014).

Vocalisations

The most common song is a variable, rapid series of 2 to 12 (often 5) loud, strident notes variously described as *kweek, kwik, kwee, quee* (Tarboton 1976) or *too, tew, tyew, kew* and *kweeah* (Short 1982). Its voice is perhaps reminiscent of that of some small falcons. It is used mainly as a territorial declaration, all year round, and is far-carrying, being audible from as far as 500 m away, and neighbouring birds will respond to each other and compete for as long as 20 minutes (Tarboton 1976). This song is generally slower than that of its

Eurasian relative but significant variations in tempo exist in both species. Like their Eurasian relative, Red-throated Wrynecks tend to sing from prominent perches such as the tops of bushes, snags and fenceposts. When calling from branches they habitually perch across them, as many passerines do. The song is performed also by mated pairs, the male's version said to be generally lower-pitched than that of his mate.

A repeated, rasping, scolding *peegh* is the most common alarm call and is given by both sexes, often when a rival is nearby. It is usually accompanied by antagonistic displays. A protracted, intense series of up to 15 *krok* notes frequently follows the *peegh* call. This, too, is usually given together with displays, often when an intruder is near, and builds up to a crescendo which can lead to an altercation. These notes are also sometimes exchanged between pair members during courtship and may then be followed by copulation. Both sexes utter soft and subdued clucking and clicking notes by the nest hole before entering it. Several *yeea* or *pyee* notes which are repeated after short pauses have also been described. Nestlings make constant wheezing squeaks until they are about six days old and later utter repeated, buzzing *tsch* sounds (Tarboton 1976).

Instrumental sounds

As is the case for the Eurasian Wryneck, Red-throated Wrynecks do not drum in the way of most woodpeckers. They do, however, tap demonstratively, usually near the nest hole. Rapid but weak tapping on branches during interactions (probably with mates), interspersed with calls and displays has also been observed (Tarboton 1976). This has been described as 'displacement behaviour' (see Chapter 10) which, if so, can nonetheless be reasonably described as a form of instrumental communication.

Distribution and range

The species has a somewhat scattered distribution in sub-Saharan Africa, with populations isolated from one another. They are resident in about 20 countries: in the north in Nigeria, Cameroon, Central African Republic, South Sudan and Ethiopia; and in the very south in Eswatini, Lesotho and South Africa. Various disjunct populations also occur in Angola, Burundi, Congo, Democratic Republic of the Congo, Gabon, Kenya, Malawi, Mozambique, Rwanda, Tanzania, Uganda and Zambia. Populations within these countries are often localised. They typically occur from around 600 m up to an elevation of

FIGURE 5.3 Range of Red-throated Wryneck (approximate).

3,300 m (in Kenya, for example, between 1,400 m and 2,500 m), but in some places in South Africa they breed near sea-level (Short 1982).

Movements

The Red-throated Wryneck is a non-migratory and sedentary species (Tarboton 2005). Claims that it does migrate are most likely unfounded, although some short-distance movements in the south of the continent have been reported. In February and March, what are thought to be 'migrants' pass through the north-east of South Africa, and occasionally South African individuals have been recorded outside their regular range. It is possible that these are local non-territorial individuals that move into an area when it becomes available, and which had previously been overlooked because they were not calling and so were inconspicuous (Tarboton 1976). Nevertheless, some populations may be nomadic, inhabiting an area for a period and then abandoning it, only to return later. There is, of course, post-breeding dispersal by recently fledged birds, and those that breed in higher mountains may make short, seasonal, altitudinal movements. More study is needed on this subject.

Status

The total world population is restricted to sub-Saharan Africa, and in most countries, perhaps surprisingly, its status is somewhat unclear. Data are lacking from many areas, with the species thought to be declining in some yet increasing in others. It is somewhat difficult to monitor as some local populations appear, disappear and then reappear in a manner that is not understood. The status of the species is best known in South Africa where it is fairly common in the east and expanding in range as the afforestation of former grasslands has created more habitat (Winkler and Christie 2002). Overall, the Red-throated Wryneck is believed to be increasing in number and is not considered to be seriously threatened. In the International Union for Conservation of Nature's (IUCN) Red List of species, it is categorised as Least Concern (BirdLife 2020).

Habitat

Typically, this is a bird of open, dry grassland with trees, especially acacia savanna and miombo woods. But they also occur in wetter, more tropical woodlands with clearings, bushy hillsides, highland grasslands with shrubs and

FIGURE 5.4 Adult Red-throated Wryneck, subspecies *aequatorialis*. November 2019, Bishangari, Ethiopia (NiB).

scattered trees, and forest edges. Red-throated Wrynecks have adapted well to habitats created by people, such as farmland with trees, suburban parks and gardens. Besides native woodlands, plantations of introduced trees such as eucalyptus and a variety of conifers are also readily occupied (Tarboton 1976).

Breeding

There are many similarities in the respective breeding behaviours of the two wryneck species, but also some significant differences. Much of what follows is based on the work of Tarboton (1976, 2005), who conducted what is still the only detailed study of the breeding biology of the Red-throated Wryneck. In common with their Palearctic relative, pairs do not excavate their own nest holes, although they may modify cavities, deepening them or enlarging the entrance slightly by pecking if the wood is soft enough. Natural tree holes and old woodpecker and barbet cavities are preferred as nesting sites, with nest boxes and hollow fenceposts also used, though there is often intense competition with Crested Barbets *Trachyphonus vaillantii*. Red-throated Wrynecks are monogamous, and males feed females during courtship. As might be expected, the timing of the breeding season varies across the continent. For example, in Cameroon it is usually from February to July, while birds have been observed breeding in April and May in Gabon and from January to May in Kenya, and in South Africa breeding occurs from August to February with egg-laying at its peak in October. One to five but typically three or four white eggs are laid. Both sexes share incubation duties for 12–15 days, and both then brood and feed their young, which fledge at 25–26 days. Two broods are regularly raised and as many as four have been recorded. Juveniles become independent around two weeks after fledging. One significant difference between the two wryneck species is that Red-throated Wrynecks are often victims of brood parasitism by honeyguides (*Indicator*), whereas Eurasian Wrynecks very rarely (if ever) suffer brood parasitism by other bird species (see Chapter 12).

Diet

Red-throated Wrynecks feed primarily on smaller terrestrial ants, which are taken in all their stages. In South Africa, Tarboton (1976) found that different ant species dominated at different times of the year and that droppings consisted almost entirely of ant remains. They also take termites and other small invertebrates, although probably only opportunistically. The same ant-based diet is fed to nestlings. Around 90% of prey is obtained from the ground (Tarboton 2005) and, when foraging there, the birds often hop with their tail cocked. They usually forage alone, probing and digging into ant mounds, and in trees they glean prey from bark, boughs and foliage rather than pecking or excavating.

Mimicry behaviour

Head-swaying and neck-twisting movements similar to those of their Eurasian relative are performed when threatened or stressed, such as when they are handled (see Chapter 10). Nestlings will use a purported 'snake-striking display' (Short 1982) in which they point the head and half-opened bill towards intruders or potential predators, before rapidly recoiling, gaping with the bill and making 'snake-like' sounds (Tarboton 1976).

Chapter 6

Communication

Wrynecks have a limited vocal repertoire. Outside the pre-breeding season they are generally silent but will sometimes sing in winter and occasionally at migration stopovers. In spring, when they have just returned to their breeding areas, both sexes are very vocal. Several perches, song posts, are often used which can give the impression that birds are ventriloquial or that there are more in an area than is actually the case. Once pairs are formed the individuals become less vocal, and any songs produced are less vociferous. After the eggs have been laid, they scale down their vocalisations further, and during incubation and when feeding small chicks, they are mostly quiet. They may, however, sing and call again just before their young fledge (Bijlsma 2014). If another Wryneck appears in an occupied territory and begins to sing, the resident pair usually resume.

Song

Wrynecks proclaim their arrival in spring by singing stridently, usually from prominent spots such as the tops of bushes, snags and fenceposts, but sometimes from more concealed perches. Some individuals do so almost incessantly from dawn to dusk, with around 40–50 bouts per hour (pers. obs.). Singing typically peaks in late morning and is more frequent on still and sunny days. Persistent advertising by means of song leads to the inspection of holes by both sexes and to courtship, which is accompanied by a range of soft, intimate calls (Ruge 1971). Individuals singing throughout spring into summer are most likely unpaired, although some may be birds that are beginning a second brood.

Wrynecks often respond strongly to playback of their song. They may sing back vigorously from afar or fly directly in and sing from an exposed perch near the sound source. Or they may fly in low and covertly, then silently

FIGURE 6.1 Wryneck singing from a favourite high song post. April 2021, Gerecse Hills, Hungary (GG).

observe the imitator from cover. As they are so responsive, researchers and ringers sometimes 'tape-lure' Wrynecks into mist-nets, particularly in migration periods (Schaub et al. 2012).

Both sexes produce a series of loud, clear, shrill, penetrating *tu*, *kew* or *quee* notes, usually 10–20, but sometimes as few as 7 and as many as 27. There are typically four notes per second, often with long pauses between phrases. The song has also been rendered as *tie-tie-tie-tie* and *kwia-kwia-kwia* (Winkler et al. 1995). The pitch of each individual note falls slightly at the end, though the entire phrase rises in pitch. Overall, the song is emphatic rather than rapid, more whining than laughing. When it begins with an acceleration and at the same time a rise in pitch, the song can be reminiscent of the vocalisations of smaller falcons such as Eurasian Kestrel *Falco tinnunculus*, Red-footed Falcon *Falco vespertinus* and Eurasian Hobby *Falco subbuteo* (Gorman 2004). A distant, fast version might suggest a Lesser Spotted Woodpecker *Dryobates minor*. Occasionally a slow version is produced, with the notes clearly separated and carefully stressed; this suggests to human ears that the bird is hesitant and reluctant to start its song. Females are sometimes said to have a hoarser version (Rasmussen and Anderton 2005); however, this needs clarification. Differences between individuals can be striking. Birds in close proximity to each other will deliver songs that vary in pitch, speed and length (both of the total song and of notes within it), and in how often they repeat their song. Finally, Wrynecks may be unique among picids in that partners will duet, softly or excitedly, responding to each other very directly and sometimes singing together by the nest hole. One will sometimes sing from the nest entrance or from inside the cavity while its mate responds from outside.

FIGURE 6.2 Wryneck singing from within the cover of a tree. June 2021, Börzsöny Hills, Hungary (GG). Note the ants, which have presumably narrowly escaped being eaten, crawling on the bird's bill and head.

FIGURE 6.3 Wryneck singing from its nest hole entrance. May 2021, Ipolydamásd, Hungary (GG).

Singing in winter

Some Wrynecks sing in their wintering quarters. This can involve single individuals or two birds singing against each other. A specific territorial winter song, which is shorter and more evenly pitched than that in spring, has been reported (Rasmussen and Anderton 2005), but again this requires clarification. It is also not always clear whether winter singing involves residents (such as in the Mediterranean), recently arrived winter visitors, individuals moving through on passage, or a combination of any of these. The function of singing in winter is likewise unclear. It may be a means of declaring a foraging area or be performed by resident individuals defending what will be their breeding area. On the other hand, some migrants sing on their wintering grounds in Africa, where they do not breed. Whatever the reason, the fact that Wrynecks sometimes respond to playback of their song in winter implies territorial behaviour (van Wijk and Tizón 2016).

Calls

In addition to the familiar song, Wrynecks produce some simple calls each of which serves a specific behavioural function. Soft *vet* and *graeb* calls are made during courtship, while contact calls between mates include soft, cooing *gruu* notes. When alarmed or disturbed, Wrynecks emit several harsh *tek, tak, tyek, tuk* or *tyuk* notes, which rise and fall in pitch. When caught (for example, for ringing) they may make muted and quiet sounds or loud and harsh rasping, grating, wheezing, scolding or screaming notes. Near the nest, especially when anxious, individuals give sharp, clicking, clucking or squealing notes, or quavering trills, and softer, slower notes recalling a Common Blackbird *Turdus merula*. Pairs by the cavity utter gentle *dyek* or *tyuk* calls in unison. During change-overs at the nest partners utter soft, quiet notes (Menzel 1968). Birds may also sing a brief, subdued version of the main song at change-overs. A scolding *sgissgissgiss* is also sometimes given at the nest entrance; Bussmann (1941) described an occasion on which a bird made this call when its mate did not appear for nest relief. Guttural sounds are sometimes uttered during copulation. If disturbed in the nest, sitting adults may make wheezing, rasping or harsh scolding sounds. Nestlings emit various squeaking, sizzling, trilling, ticking and tinkling sounds, the tinkling calls recalling those of some *Locustella* warblers (pers. obs.). All of these calls can be loud or muted. As they grow, they also beg for food with rapid rasping noises which intensify from about day 10 and from about 17 days they may make hissing sounds (Bussmann 1941). Just before fledging well-grown young also beg with ticking calls at the nest entrance and continue to do so when outside. Wrynecks on migration are seldom vocal, although an interesting report concerned an individual on passage in September near Dover, England, which was heard making a short, soft *z-e-e-p* trill which had a thrush-like quality. It was concluded that this may have been a young bird that had fledged quite recently, and which was continuing to utter its begging calls (Chantler 1991).

Instrumental sounds

Wrynecks also communicate by means of instrumental (non-vocal) signals. For instance, there exist two basic forms of tapping rhythms, which correspond to different behaviours. Most are fast but rather muted and are used in communication between members of a breeding pair, while others are very brief and slow and seemingly not used between pair members (Turner and Gorman 2021). Fast tapping is done as part of courtship and is performed at any time from the first copulation through to hatching of the eggs, and it occurs also inside nest holes. The slower form, given in short bursts, is similar to the 'nest-showing' signals of the pied woodpeckers (Winkler and Short 1978). Tapping is done when a potential nesting site is being advertised by a bird to its partner, to confirm its ownership and perhaps to enforce the pair bond. It is performed also during incubation and brooding change-overs (Menzel 1968), when pairs will excitedly and energetically peck and rhythmically tap around the cavity entrance, producing a short, soft series of knocks or a double tap. Double taps may also be given by an adult Wryneck on exiting from the hole after feeding newly hatched young. Triple taps in immediate response to a mate's song are sometimes delivered. Faster bursts may be made deep within the nest chamber, with one or both adults present inside. Although seldom heard, mated pairs will even indulge in tapping duets, one bird tapping at the nest hole entrance and its partner inside responding with more rapid taps. Whatever form they take, such sounds are not the same as those made by pecking when adults modify a cavity or when nestlings probe the walls of the chamber. Sounds resulting from these activities are typically erratic in both rhythm and strength, and sound more like scratching or tearing; they are not instrumental signals. Very occasionally, a Wryneck may tap weakly on branches away from the nest site (Gorman 2004). The function of this behaviour is unclear; it may be a form of communication, but also perhaps linked to bill-cleaning, foraging, or is possibly displacement behaviour (see Chapter 10). When agitated near the nest, for instance when a predator is close by, Wrynecks will also fly back and forth with rapid, exaggerated flapping of the wings to produce a whirring sound (pers. obs.). This is likely to be an attempt to intimidate or at least distract an intruder.

Drumming or tapping?

The terminology used by some authors to define the instrumental sounds that Wrynecks produce has occasionally been ambiguous. Some have described these acoustic sounds as drumming (Short 1982; Glutz and Bauer 1994). Referencing Witherby et al. (1938), Short wrote that Wrynecks will drum 'woodpeckerlike, but weakly, slowly, and not audible at any distance' and Puhlmann (1914) describes soft drumming as being done inside the cavity during nest preparation. More specifically, Schneider (1961) outlined four rolls, ranging from six to ten strikes, given by a bird at the entrance to a nest box in response to the arrival of a second bird.

Although they may develop rhythmically, the instrumental sounds which Wrynecks produce are invariably soft and are never prolonged, nor are they as regular as the drumming performed by many other picids. While they are capable of producing drum-rolls of sorts, they do not do so in the functional, territorial sense commonly associated with the term. Wrynecks do not produce distinct rapid series of similar strikes that can realistically be defined as drumming rolls. Hence, their instrumental sounds are probably best described as 'demonstrative tapping' or 'rhythmic tapping' (Ruge et al. 1988). Indeed, their fine-tipped, weak bill and the less developed associated muscles compared with those of other woodpeckers mean that Wrynecks are in fact unable to drum in the true sense, that is, to advertise themselves loudly and powerfully.

It is perhaps interesting to speculate on the similarities between Wrynecks and the true woodpeckers regarding their evolutionary divergence. The theory that instrumental signals, including eventually drumming, may have evolved from the action of excavation (Zabka 1980) would suggest that separation occurred after the development of a strong bill and associated muscles. In this case, Wrynecks would be evolving away from a common, strong-billed ancestor. Nonetheless, many similarities to other woodpeckers remain. For example, some of the Melanerpine subfamily of the Americas use mutual tapping, with one bird inside the cavity and one outside (Winkler and Christie 2002).

Spectrograms

The vocalisations of birds can be illustrated in spectrograms, which are visual depictions (graphs) of recorded sounds. They provide an alternative to using mnemonics and phonetics, which are rather subjective as a means of describing vocalisations. Measurements of call parameters, such as of cadence, can be taken from spectrograms and this helps to distinguish patterns in the structures of the recorded sounds. In the examples below, the frequency of a sound is plotted against time, with time shown on the horizontal x-axis in seconds and pitch (frequency) in kilohertz (kHz) on the vertical y-axis. In all cases the sex of the bird was undetermined.

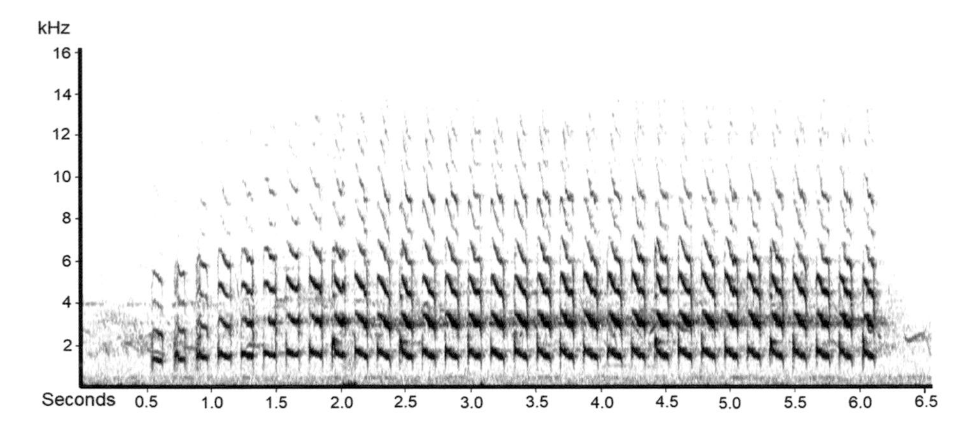

FIGURE 6.4 Wryneck song. The bird was singing from a snag in a row of trees in farmland. The song is a series of repeated, loud notes with somewhat harsh-toned elements, often described as falcon-like. In this sequence the individual elements are comprised of multiple harmonics. Each element quickly rises to its highest frequency before falling to its lowest, as shown by the downward slope from left to right. Recorded by Gerard Gorman, Gerecse Hills, Hungary, April 2021.

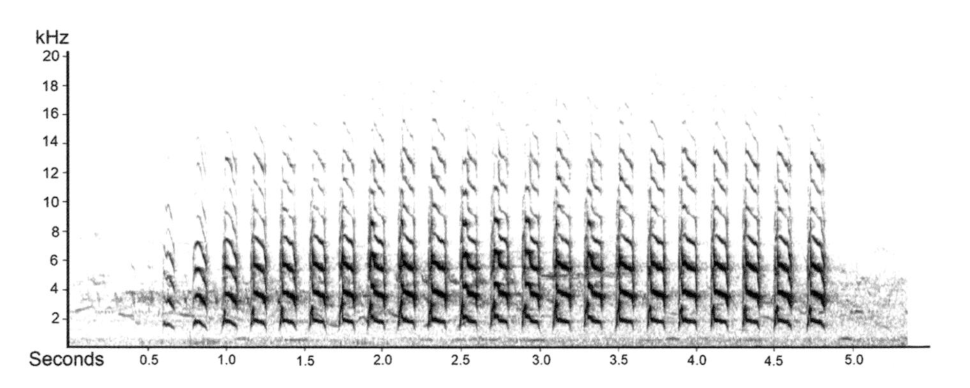

FIGURE 6.5 Wryneck song. The bird was calling from within thorny scrub. The song is a series of repeated, loud notes with somewhat harsh-toned elements, often described as falcon-like. In this sequence there are 22 individual elements and it can be seen that these comprise multiple harmonics with most energy within the first three elements. Each element quickly rises to its highest frequency before falling to its lowest, as shown by the downward slope from left to right. Recorded by Daniel Alder, Bükk Hills, Hungary, May 2002.

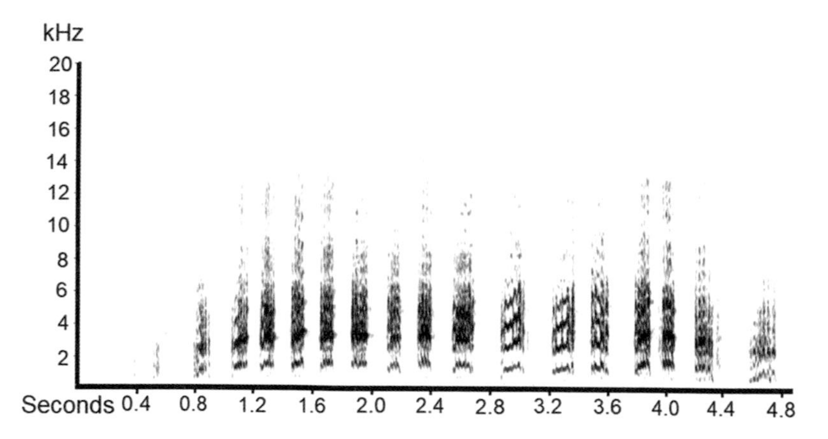

FIGURE 6.6 Wryneck call. This call was given by one adult bird in the presence of another during nest relief and appears to convey agitated behaviour in the absence of a response. This call is more complex in structure than that shown in 6.5, with multiple harmonics and frequency modulations which show up in the zigzag pattern across each harmonic. The sound is richer with a harshness and intensity to its tonality. Recorded by Kyle Turner, Bükk Hills, Hungary, June 2007.

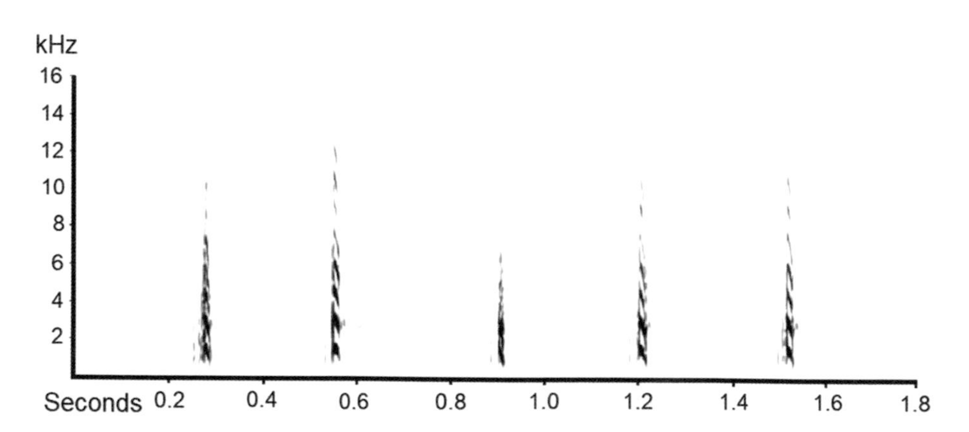

FIGURE 6.7 Wryneck alarm call. This is a short extract from a long series of rising and falling call elements given in alarm by an adult close to the nest in the presence of an inquisitive Starling *Sturnus vulgaris*. The call comprises multiple harmonics which start with the first harmonic at around 1.2 kHz. Most of the energy of the call is within the first four harmonics. Recorded by Kyle Turner, Bükk Hills, Hungary, June 2007.

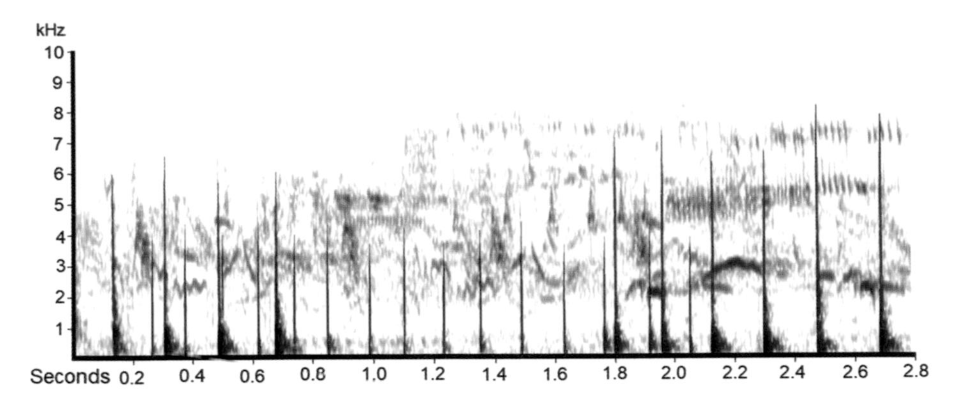

FIGURE 6.8 Wryneck tapping. This is an adult bird tapping at a nesting cavity entrance. It produced two bursts of four and six strikes each (coloured blue) while the mate replied with faster taps (coloured red) from inside the cavity chamber. Recorded by Kyle Turner, Bükk Hills, Hungary, June 2007.

Chapter 7

Distribution, Trends and Status

The Wryneck's global distribution covers parts of Europe, the Middle East, Asia and Africa. The breeding range lies within the Palearctic, approximately between latitudes 35° and 64° N, from the Atlantic coast of Europe and a small area in North Africa to the Asian Pacific coast and Japan and is estimated to cover 38,400,000 km^2. The winter range extends southwards to around the Equator and has been projected at 56,200,000 km^2 (BirdLife 2021).

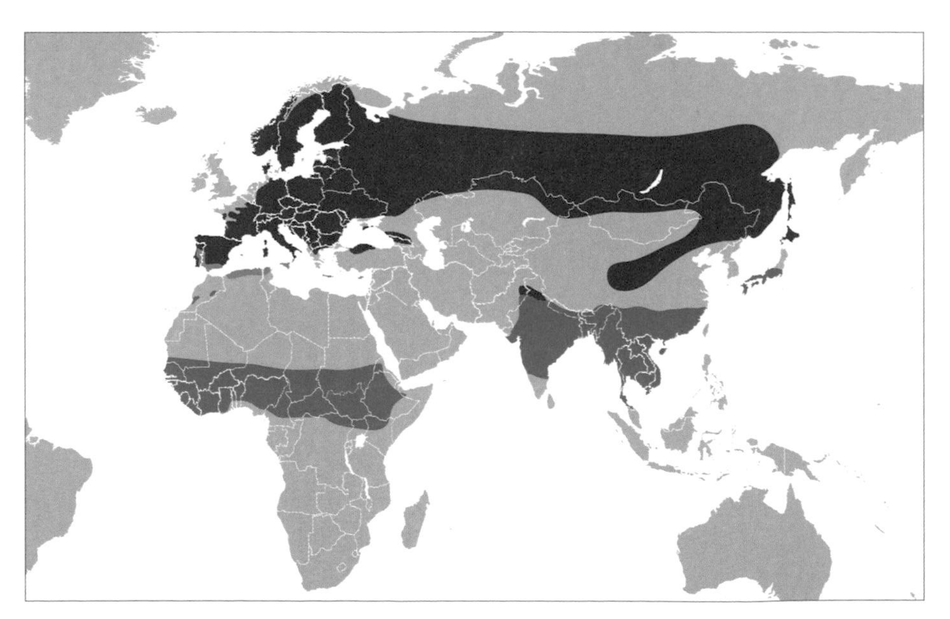

FIGURE 7.1 Global distribution of Eurasian Wryneck. Ranges (approximate): breeding – red; resident – green; blue – winter.

Elevation

Wrynecks are found mostly in lowland and hilly areas, from sea-level up to 1,000 m, though some occur at quite high elevations on sunny slopes. In the Alps and Apennines, they are regular up to 1,200 m, occasionally nesting as high as 2,100 m (Brichetti and Fracasso 2020). In the Romanian Carpathians they breed up to 1,300 m (Petrovici et al. 2015), in the Caucasus to 1,600 m and in Kazakhstan at 1,900 m, although they have been recorded at 2,800 m in the Altai (Wassink 2015). In the western Himalayas they breed at 1,500–3,300 m, with passage birds seen in Bhutan and the eastern Himalayas at 3,800 m (Grimmett et al. 2019). In African wintering areas they are found up to 2,500 m (Short 1988a). They typically occur to 1,800 m in South-East Asia (Winkler et al. 2020).

Population

The overall global population has been estimated at between 3,000,000 and 7,100,000 individuals, with a European population at 674,000 to 1,600,000 pairs (BirdLife 2021). In terms of breeding in Europe, Wrynecks undergo greater fluctuations in numbers and occurrence than any other woodpecker. The most likely reason for this is that, unlike other European picids, the bulk of the population is migratory which means conditions at passage stopovers and in wintering areas influence numbers. Although there is less information from Asia, this may also be the case in that region.

European distribution

The European breeding range stretches from sub-arctic Fennoscandia to the Mediterranean and Black seas and is estimated to comprise 45% of the global range (BirdLife 2021). A temperate, drier, continental climate seems to be preferred, which may be why the species is less common along the Atlantic coast (Vogel 1997). Overall, there has been a significant reduction in range during the twentieth century (Gregory et al. 2007; Keller et al. 2020). Wrynecks have probably never bred in Ireland and have been lost as breeding birds from Britain.

Wintering in Europe

Most of Europe's population winters in sub-Saharan Africa but some remain in the Mediterranean basin (Reichlin et al. 2009). Although the Wryneck has often been regarded as scarce in winter in Iberia, it is probably not uncommon, particularly in the south-west, along the Mediterranean coast and in the Balearic Islands which are all warm locations where ants are numerous even in winter (Pinilla 2021). In Spain, Wrynecks occur nationwide from November to February, mainly in Andalusia and Extremadura, but also as far north as Asturias (González et al. 2002; Carbonell 2012). Around 200 winter in Catalonia every year (Herrando et al. 2011). In Italy, most winter in the south but also further north in Tuscany, and in milder years as far north as the Po valley (Brichetti and Fracasso 2020). Ringing data suggests there is fidelity to wintering sites. Several Wrynecks caught in autumn on Vivara in the Gulf of Naples were subsequently re-trapped, suggesting they remain on the island. About 26% of winter-ringed birds were recaptured the following year and some returned when in their third year (Scebba and Lövei 1985). Yet others ringed in early winter apparently did not stay on Vivara for the whole season but disappeared after one or two months, and some ringed in February lingered until spring. It was thought that these were the birds most likely to return the following year. Since the 1950s, and especially the 1980s, increasing numbers have wintered in southern France (Orsini 1997; Issa and Muller 2015). Others occur in Malta, mainland Greece and islands such as Crete and Cyprus (Richardson and Porter 2020). There are even occasional winter records from Britain (Booth et al. 1968; Donovan 1968). Ultimately, the number of Wrynecks wintering in Europe fluctuates annually and probably depends on temperature and weather conditions.

Residents or visitors?

Whether Wrynecks wintering in Europe are local residents, migrants from elsewhere or a combination of both is hard to assess. In Corsica, Sardinia and Sicily, for example, some may be residents rather than migrants from the mainland. A study of 496 ringed birds (94 re-trapped) from Spain's Mallorca, Sa Dragonera and Cabrera islands, confirmed a breeding population, but it was unclear whether birds seen there in winter were from that population, and therefore residents, or visitors (Garcías 2010). It is known from Swiss and German breeding birds fitted with geolocators that some winter in Morocco and Iberia (van Wijk et al. 2013). Another way to determine whether wintering birds are permanent residents or not can be to take wing measurements. Wrynecks from northern Europe tend to be longer-winged than southern birds (Scebba and Lövei 1985).

European status

Specific information on trends is only known from a few countries, but overall the population is considered to be in decline. This has been especially sharp in the west and centre of the continent where numbers crashed from 1980 (Gregory et al. 2007). The situation in the east, in the Balkans, Belarus, Russia and Ukraine, is considered healthier (Winkler et al. 2020), with most of Europe's pairs believed to occur there (Keller et al. 2020). A demographic study in Switzerland revealed that juvenile survival and immigration from other populations greatly influenced local numbers. Population trends are therefore influenced by dispersal from, and immigration into, each area where the species breeds and appear to be synchronised across different countries (Schaub et al. 2012.

Selected European countries

Armenia: Rare breeder (Adamian and Klem 1999). Perhaps under recorded.

Austria: Declined sharply since the 1950s, especially in the north. Long-term trend (1998–2020) stable, short term (2015–20) moderate increase (Teufelbauer and Seaman 2021) to 3,300–4,900 pairs (Dvorak 2019).

Belgium: No longer breeds in the north (Flanders) (Vermeersch et al. 2020). In the south (Wallonia) 45–58 pairs in 2007 (Jacobt et al. 2010), now 66–101 pairs.

Britain: No longer breeds. Occurs on passage, some in spring but most in autumn. For example, in the period 1986–2003, around 300 a year were reported (Fraser and Rogers 2006).

Bulgaria: Breeds nationwide but distribution patchy. Absent from high mountains and intensively farmed areas (Iankov 2007); 2,500–5,000 pairs but declining.

Croatia: Common and widespread, most along coast. Population size and trends unknown.

Cyprus: Does not breed. Common on passage in spring, fewer in autumn, rare in winter (Richardson and Porter 2020).

Czech Republic: After a decline now stable at 2,500–5,000 pairs (Šťastný and Bejček 2021).

Denmark: Long-term decline, but recent slight increase as it has adapted to habitats where it previously did not breed, such as conifer plantations with clearings on dunes (Andersen 2018); 212–50 pairs, stronghold Jutland (DOF 2020).

Estonia: Widespread on mainland and larger islands, but sharp decline recently observed; 5,000–10,000 pairs in 2013–17 (Elts et al. 2019).

Finland: Mainly in the south and centre of the country, rarer in Lapland. Continuous decline between 1952 and 1977 (Linkola 1978), strong decline of around 75% in the 1980s and 1990s, then more or less stable in the 2000s with 10,000–20,000 pairs (Valkama et al. 2011).

France: Fairly widespread in the east, south and on Corsica, scarce in the west and north along the Atlantic coast (Keller et al. 2020). Overall, after a strong decline, now considered stable with 20,000–40,000 pairs. Some winter, mainly along the Mediterranean coast (Orsini 1997; Issa and Muller 2015).

Germany: Mainly found in the east, middle and south of the country, with highest densities in the north-east. Formerly widespread, strong declines noted from the late twentieth century; 8,500–15,500 pairs (Bairlein et al. 2014; Gedeon et al. 2014).

Gibraltar: Only on passage, about 50 records a year.

Greece: Concentrated in the north, perhaps more common than thought in the south, but unclear if birds there in spring are breeders or on passage (Keller et al. 2020); 300–500 pairs.

Hungary: Fairly widespread. Undergoes large fluctuations, but overall slightly increased in 2014–18; 18,000–25,000 pairs (Gorman et al. 2021).

Ireland: Does not breed but occurs annually, mainly in autumn. Most on the south coast, extremely rare in the north (Robinson et al. 2020).

Italy: Common, widespread breeder on the peninsula, some on Sardinia, Sicily and other islands, although gaps in distribution exist, especially in the south in Apulia, Campania and Calabria. Some residents, some partially migratory and large numbers on passage. Trends vary regionally but since 2000 the overall range has reduced and fragmented, particularly in Tuscany, Po Plain and Sardinia; 50,000–100,000 pairs in the early 2000s, more recently 40,000–80,000 (Spina and Volponi 2008; Brichetti and Fracasso 2020).

Latvia: Widespread, having increased in range since 2000. Current trend stable; 4,000–10,000 pairs (Ķerus et al. 2021).

FIGURE 7.2 Wrynecks do not nest on Malta but stop off when migrating in both spring and autumn. This one was there in May 2021 (RV).

Lithuania: Currently 10,000–12,000 pairs but decreasing.

Luxembourg: Declining, 50–100 pairs.

Malta: Does not breed, but common on passage and some overwinter. Recorded in all months except June and July, most in September.

Netherlands: Declined sharply, now seriously threatened, though slight recent increase noted. Resident, 95–125 pairs in 2019, most inland; 500–2,000 on passage each year, mostly on coast (Sovon 2021).

Norway: Trends unclear, perhaps some local increases but overall in decline. Most common in the south but occurs as far north as Finnmark; 1,500–3,000 pairs (Shimmings and Øien 2015).

Poland: Fairly common and widespread; 38,000–64,000 pairs in 2008–12 (Chodkiewicz et al. 2015), recent increase noted.

Portugal: Some probably resident in the south, others pass through on migration; 1,000–5,000 pairs, but overall trends unknown.

Romania: Despite a sharp decline first noted in the 1960s, still widespread with 30,000–70,000 pairs (Petrovici et al. 2015).

Russia: Common and widespread; 300,000–730,000 pairs in European Russia (Mischenko 2017).

Serbia: Widespread, stable population of 4,700–6,800 pairs in the period 2008–13 (Puzović et al. 2015).

Slovakia: Fairly common and widespread, now stable after long-term decline; 2,500–4,000 pairs estimated previously (Krištín et al. 2014).

Slovenia: Common and widespread in hilly areas, farmland and riparian woods, avoiding high Alps and dense forests; population stable at 5,000–10,000 pairs (Mihelič et al. 2019).

Spain: Widespread, has expanded range. Common in the north (except for the north-west), east, south-west and Mallorca (González et al. 2002). Some overwinter, mainly in the south. Estimate of 10,000 pairs in 2003 (Martí and Del Moral 2003) but no recent estimate.

Sweden: Breeding population in the late 1990s considered to be just 20% of that in the late 1970s and early 1980s (Svensson 2000). There was an upward trend in the early 2000s but since 2010 the trend has been downward again; 25,000 pairs in 2018 (Wirdheim 2020).

Switzerland: Declined, even disappeared, from many areas since the end of the twentieth century, only occurring regularly in the Alps and the south (Weisshaupt et al. 2011). A slight increase to 1,000–2,500 pairs was reported by Knaus et al. (2020).

The Middle East

Wrynecks occur widely on migration throughout the Middle East and can occur almost anywhere as there are abundant suitable stopover sites such as scrub, groves, orchards, vineyards, wadis, and wooded rural and urban settlements. The species is almost certainly under-recorded owing to the vastness of the region, its usually discreet behaviour when on passage and because there are relatively few observers. Almost no local studies have focused on the species and information is lacking from many countries such as Iraq, Kuwait, Saudi Arabia, Qatar and Yemen. Although essentially a passage migrant, unknown but presumably small numbers overwinter. Of the countries listed, Wrynecks only breed in Turkey and possibly northern Iran.

Selected Middle Eastern countries

Bahrain: Scarce passage migrant and winter visitor, mainly from September to April (Hirschfeld 1995).

Iran: Probably widespread on passage. Some winter in the south-east in Sistan-Baluchestan and Hormozgan. Suspected to breed in the north in Gilan, Ardebil and East Azerbaijan (Sehhatisabet et al. 2006; Tohidifar 2008).

Israel: Common on passage, mainly March to April and September, but on average only 120 ringed annually. Scarce in winter, most in the north (Yoav Perlman pers. comm.).

Jordan: Passage migrant but data lacking (Stêpniewska et al. 2011).

Lebanon: Scarce to uncommon passage migrant (Ramadan-Jaradi et al. 2020).

Oman: Some overwinter from late August to late April; fairly common on passage (Eriksen and Porter 2018).

Syria: Uncommon on passage (Murdoch and Betton 2008).

Turkey: Common on passage, some breed and a few overwinter (Kirwan et al. 1999; Keller et al. 2020).

United Arab Emirates: Regular passage migrant, irregular in winter (Richardson 1990).

Asia

In many parts of this vast continent almost nothing is known about the status of Wrynecks. Data on numbers and trends are limited as few studies have been done and, in many places, there is a distinct lack of observers. For countries such as Kyrgyzstan, Tajikistan, Turkmenistan and Uzbekistan, all that is known is that birds move through on passage (Ayé et al. 2012). In Siberia, Asian Russia and much of the Far East, for example, in North Korea, breeding status is largely a mystery. Wintering ranges, too, are often unclear due to the fact that the species is mostly solitary, secretive and silent in that season. Breeding populations from western Asia are thought to winter in East Africa, and eastern Asian populations south of the Himalayas on the Indian Subcontinent and in Indo-China (Rasmussen and Anderton 2005; Winkler et al. 2020). Birds from northern Japan, the Kurile Islands and Hokkaido (and probably the Sakhalin Islands) migrate south to Honshu and Kyushu (Winkler et al. 2015). Wrynecks that winter in the Middle East and south-east Iran may also be from western Asia (Sehhatisabet et al. 2006).

Selected Asian countries

Afghanistan: Status and distribution uncertain; probably only breeds in the extreme east, elsewhere on passage (Ayé et al. 2012). Unclear how many observed are breeders or migrants (Rasmussen and Anderton 2005).

Bangladesh: Does not breed. Uncommon winter visitor, arriving late September, departing early April (Chowdhury 2020).

Bhutan: Passage migrant, some winter in foothills and lowlands (Grimmett et al. 2019).

China: Locally common, breeding in the west, north and north-east, winters in the south and on Hainan and Taiwan, others on passage (MacKinnon and Phillipps 2000; Robson 2011).

Hong Kong: Uncommon on passage, rare winter visitor (3–12 birds annually) occurring from August to April.

India: Breeds in the extreme north-west in the Himalayas. Widespread in winter in lowlands across the north and middle of the subcontinent (Rasmussen and Anderton 2005; Grimmett et al. 2011).

Japan: Uncommon breeder on Hokkaido and northern Honshu; winters to the south, from central Honshu to Kyushu (Chikara 2019).

Kazakhstan: Common breeder and passage migrant nationwide (Wassink 2015).

Laos: Winters in the north and centre of the country, scarce in the south (Robson 2011).

Malaysia: Rare vagrant to Peninsular Malaysia (MNS 2015).

Mongolia: Breeds mainly in the north; probably many on passage but data lacking.

Myanmar: Winter visitor occurring from October to April (Robson 2011; Winkler et al. 2020).

Nepal: Regular on passage and some winter at lower elevations.

Pakistan: On passage in Lower Sind and Balochistan; winter visitor to the Punjab, some breed in the north (Grimmett et al. 2008).

South Korea: Does not breed. Mainly on passage, rare in winter.

Thailand: Widespread and fairly common in winter, particularly in the north, north-east and centre of the country, as far south as Phetchaburi province.

Vietnam: Scarce on passage and in winter, mainly in west and east Tonkin, northern Annam and Cochinchina.

FIGURE 7.3 Wrynecks do not breed in Nepal but stop off on migration and some overwinter. This one was photographed in February 2008 in Koshi tappu (NeB).

Africa

Breeding has only been confirmed in two countries, Algeria and Tunisia, where there are resident populations (Short 1988a). In Tunisia, however, it is only found in the very north-west and its overall status is uncertain as others pass through on migration (Isenmann et al. 2005). Most birds in Africa are winter visitors from European (van Wijk et al. 2013) and west Asian breeding populations. They occur from 13° N southwards in the Guinea savanna and arid regions of the adjacent Sudanian vegetation zone, in a wide band from Senegal, Gambia, Guinea-Bissau, Sierra Leone and Liberia in the west, to Eritrea and Ethiopia in the east. They inhabit acacia savanna and broad-leaved woodland, rainforest edges and clearings, and also cultivated land (Short 1988a). The more forested areas of the Guineo-Congolian region are avoided. In West Africa they are uncommon on the coast, occurring rather in the interior. In East Africa they occur mainly in the Somali-Masai zone. Wrynecks are exceptionally rare south of the Equator, the most southerly record documented at the time of writing being a bird photographed at Nguuni, near Mombasa, Kenya, on 20 April 2012 (Pearson and Jackson 2016). Interestingly, this bird was also an extremely late record for the region. Most Wrynecks are back in their breeding areas by that date, which might suggest that some unobtrusively remain in Africa rather than returning north to breed.

In many countries where they winter, Wrynecks are probably overlooked as suitable habitat is abundant, and birds are often silent and inconspicuous. Hence, they are often said to be scarce or rare in the literature, but are probably more numerous than documented. Information from countries such as Burkino Faso, Egypt, Gabon, Guinea-Bissau, Guinea, Ivory Coast, Libya, Mali, Sierra Leone, Somalia and Western Sahara is particularly scant. Wrynecks pass though these countries on migration, and some may overwinter, but numbers are unknown. The species is a vagrant visitor to the Central African Republic, Congo, Democratic Republic of the Congo and Tanzania. Some fidelity to winter sites has been noted. For example, in ringing expeditions to the lower Senegal River Valley between 1987 and 1993, one individual was re-trapped at the same site in three different winters (Sauvage et al. 1998).

Selected African countries

Algeria: Uncommon breeder. Residents nest below 500 m, birds occurring at higher elevations, to 2,500 m or more, thought to be winter visitors from Europe (Short 1988a; Isenmann and Moali 2000).

Benin: Scarce from December to March (Dowsett-Lemaire and Dowsett 2019).

FIGURE 7.4 Stony semi-desert in Western Sahara (SA). Unknown numbers of Wrynecks pass through vast areas of such terrain in Africa in both spring and autumn.

Cameroon: Scarce from December to April (Languy 2019).

Chad: Scarce in winter, extreme dates 3 September and 23 April (Salvan 1967–9).

Djibouti: Fairly common on passage and in winter (Redman et al. 2011).

Eritrea: Fairly common on passage in both spring and autumn. Some in winter (Ash and Atkins 2009).

Ethiopia: Fairly common on passage in both spring and autumn. Some in winter (Ash and Atkins 2009).

Gambia: Uncommon visitor from November to April (Morel and Morel 1990; Borrow and Demey 2011).

Ghana: Widespread from late October to late April (Dowsett-Lemaire and Dowsett 2014).

Kenya: Scarce visitor (Stevenson and Fanshawe 2002), for example, individuals ringed in the Tsavo West National Park on 5 January 2000 and 1 December 2003 (Kennedy 2019). Of 22 dated records, extremes are 9 October and 20 April (reports in *Scopus*; Robert Dowsett and Françoise Dowsett-Lemaire pers. comm.).

Liberia: Scarce in winter in northern savannas, rarer on coast (Gatter 1997).

Mauritania: Mainly on passage, some in winter (Isenmann et al. 2010).

Morocco: Occurs on passage, some in winter (Bergier et al. 2017).

Niger: Scarce, from September to March (Villiers 1950).

Nigeria: Ringing projects have found birds to be frequent on passage in the north-east by Lake Chad with some overwintering (Gustafsson et al. 2021); extreme dates 1 September (Dowsett 1969) and 2 May (Dowsett 1968).

Senegal: Occurs from November to April (Morel and Morel 1990; Borrow and Demey 2011); earliest recorded date 13 October (Rodwell et al. 1996).

Somalia: Rare vagrant (Redman et al. 2011).

Sudan and South Sudan: In the past said to be uncommon in winter, from October to March (Cave and Macdonald 1955). Current status unknown.

Tanzania: Exceptionally rare visitor with, at the time of writing, only two confirmed records, both in the very north, the first on 11 January 2010 in the Arusha National Park (Kennedy 2019).

Togo: Scarce visitor, December to March (Dowsett-Lemaire and Dowsett 2019).

Tunisia: Rare resident, breeds in the north-west. Others on passage in both spring and autumn (Isenmann et al. 2005).

Uganda: Scarce visitor (Stevenson and Fanshawe 2002); extreme dates 21 October and 6 March (Pearson and Turner 1986).

Overall trends

Wrynecks are known to experience unpredictable fluctuations in breeding numbers. Sudden peaks may be followed by equally abrupt slumps (Glutz and Bauer 1994). Nevertheless, across much of Europe where they were once common, breeding populations have declined. Ringing data, too, points to this. To give one example, records show that numbers on passage in Schleswig-Holstein, Germany, fell dramatically in the period 1965–98 (Busche 2004). This trend has been documented across the whole of the continent and seems to have begun early in the twentieth century, and possibly before that. A detailed long-term (1980–2009) assessment of the population trends of some European Afro-Palearctic migrant birds, which combined national breeding bird survey data from across Europe, found Wrynecks had experienced a decline of 49% (Vickery et al. 2014). For comparison, two often sympatric species in European breeding areas, the Barred Warbler and Red-backed Shrike, experienced declines of 48% and 36%, respectively. Another estimate put the Wryneck's rate of decline at 57% for the period 1980 to 2013 (Sanderson et al. 2006). Yet, there is a marked contrast in trends across Europe. In the west, declines documented in the 1980s have continued, but after a similar situation in central Europe numbers seemed to stabilise at the end of the twentieth century (Keller et al. 2020). In the same periods large populations in the east were presumed to have been more resilient, but it is possible that a lack of quantitative data may have hidden negative trends (Zwarts et al. 2009).

It is generally agreed that European populations of all long-distance migrant birds have on average suffered more than short-distance ones, which have been largely stable or have increased (Gregory et al. 2007). The main reasons for the declines of some species are considered to be loss of suitable habitat and intensive farming, including increased cultivation and changing forestry practices (BirdLife 2021). In Africa, land-use, deforestation and drought are suspected (Zwarts et al. 2009). Occupying a wide breeding range across Europe, Wrynecks are both long- and short-distance migrants, and some are residents. Hence, several factors, including those beyond breeding areas, are probably negatively affecting the species (for more on causes and threats see Chapter 9). Fortunately, not all trends are negative. When local conditions change Wrynecks can return. An example is Baden-Wurttemberg in south-west Germany where numbers fell sharply before the millennium. Then, from around 2004, observations significantly increased in areas of the Black Forest which had been opened up by storm damage (particularly a major event in 1999 that felled many trees). Subsequently Wrynecks re-established themselves in the new semi-open habitats where abundant standing and lying deadwood benefited ants (del Val et al. 2018).

Extirpation – the case of Britain

The Wryneck population in Britain has probably always been somewhat peripheral owing to geographical location and a relatively wet oceanic climate. Historically, however, the species was once fairly common and widespread, recorded in 54 counties between 1875 and 1900, although not recorded breeding in Scotland (Holloway 1996). At the end of the 1800s and during the 1900s, numbers began to fall and the range contracted considerably. This began in the north around 1830 and by 1905 Wrynecks had disappeared from Wales (Lovegrove et al. 1994). By the mid-1950s, they were confined to central and south-east England, but only numerous in Surrey and Kent (Monk 1963). But by the 1960s they had retracted to even smaller isolated populations, mostly in the very south-east (Peal 1968). The decline continued until

FIGURE 7.5 Today Wrynecks only occur in Britain on migration. This individual stayed for a few days in Hampshire, England, in September 2015 (RD).

the last confirmed nesting in England, in Buckinghamshire, in 1985 (Holling et al. 2012).

Despite this, through the 1950s there had been reports of singing Wrynecks in the Scottish Highlands, particularly to the west of the Cairngorms. Subsequently, in 1969 the first three breeding records were confirmed (Burton et al. 1970). One pair nested in a hollow in a Rowan *Sorbus aucuparia* and two in old Great Spotted Woodpecker holes. It was thought that this would herald a return of the Wryneck to Britain as a whole, and at the end of the twentieth century some pairs still bred in Scotland – in Deeside, Badenoch and Strathspey – nesting in open natural and semi-natural coniferous woodlands, in habitats not unlike those in Fennoscandia. But numbers were never large; for example, in Strathspey they peaked at ten pairs in the late 1970s and 1980s, and the population subsequently dwindled. The last confirmed breeding pair in Ross-shire in the Scottish Highlands (and all Britain) was in 2002 when a nest with young was found (Forrester and Andrews 2007). Singing birds seemingly on territory have been reported at several sites in the ensuing years (Holling et al. 2012), but since 2013 there have been no verified breeding records (Eaton and Holling 2020). Today, the species is officially classed as a former breeder. It is therefore probably the first breeding species to have been lost from the country in modern times and, regrettably, Britain has the ignominy of being the only country in Europe from which Wrynecks have become extirpated as breeding birds. Today they are only regularly seen when on through passage (Eaton et al. 2015) but even these birds are fewer in number than in earlier times. For example, only 0–4 are caught each year at the Fair Isle Bird Observatory in the North Sea, usually in May and September. Many more used to visit this island, 'scores along roads' reported in 1943 (Dymond 1991).

The Wryneck's disappearance in Britain has not been fully explained, but it is presumed that the factors that have resulted in it decreasing in other parts of Europe in recent times (habitat loss and degradation) were the cause. Yet the declining trend started in the 1800s, before the intensive farming of modern times began, and therefore problems with food resources have also been suspected (Taylor 1993). It is noteworthy, too, that declines were also recorded in Germany as long ago as the late 1800s (Vogel 1997). In the Scottish context, to this day the formerly used breeding area still exists and has not appreciably changed, though some dry areas are now boggy owing to habitat management. Habitat loss and degradation could not have directly affected the Scottish breeders as they formerly nested in a non-agricultural area. Furthermore, in many places in England there remains suitable habitat such as parkland, old orchards and large gardens with cavity trees.

Chapter 8

Habitats

The environments that Wrynecks inhabit vary according to region and season. The huge range of the species takes in four eco-climate zones (boreal, temperate, subtropical, tropical) and they breed from sea-level to sub-Alpine elevations, which results in considerable diversity in the habitats they use. For example, in the Mediterranean they breed in almond and olive groves, in Fennoscandia in stands of birch (*Betula*) and pine (*Pinus*) and in central Europe often in orchards. On migration, they use an even wider range of habitats and

FIGURE 8.1 Mosaic habitats that comprise traditionally farmed pastures, meadows, patches of rocky or barren ground and trees with cavities are ideal for Wrynecks. Börzsöny Hills, Hungary (GG).

wintering areas; some, such as wooded savanna in Africa, are quite different to their breeding habitats. Nonetheless, Wrynecks are ultimately a woodland-grassland species, in all seasons essentially requiring warm, open landscapes (avoiding continuous and dense tracts of forest) where ground-living ants, their favoured food (see Chapter 14), are abundant (Coudrain et al. 2010).

Breeding habitats

The species typically breeds in semi-open, drier, sunlit wooded areas with sparse ground cover (Freitag 1996). Across their Eurasian range they nest in a wide range of habitats, which can differ significantly in character, such as woodland with adjacent pastures, meadows, fallow land, dunes, heaths, parkland, orchards, olive and chestnut groves, vineyards, pollarded hedgerows, open pine forests, the edges of plantations, and even rural and suburban gardens. In all areas two main requirements need to be met in the breeding season: open bare ground or short grass areas with high densities of ants, and suitable cavities for nesting (Glutz and Bauer 1994). Breeding habitat occupancy is known to diminish with increased cultivation and landscape uniformity (Mermod et al. 2009).

FIGURE 8.2 In many parts of Europe traditionally grazed upland pastures with old trees are classic breeding areas for Wrynecks. Asturias, Spain (JMFDF).

FIGURE 8.3 A rural garden with old fruit trees with cavities, patches of grazed and bare ground and a distinct lack of pesticide use. Wrynecks and other birds often do well in habitats managed in such a way. Ipolydamásd, Hungary (GG).

Woodlands

Depending on the region, Wrynecks are found in deciduous, mixed conifer and deciduous and, to a lesser extent, pure conifer stands. In Spain they often nest in Cork Oak *Quercus suber* woods, riverine woodlands, mature tamarisk stands and plantations of poplar (*Populus*) and especially eucalyptus (Martínez and Domínguez-Santaella 1997). At northern latitudes and in mountains, pine and larch (*Larix*) stands are used (Winkler et al. 2020). In North Africa they breed in various woodlands, especially of oak but also in alder (*Alnus*) and eucalyptus, and locally in parks and gardens (Short 1988a). In Japan they also nest in more humid woodlands (Yoshimura et al. 2003). Conifer plantations often host ant colonies within clear-cut areas, especially at their edges, but Wrynecks only use them for breeding when snags with cavities are present and these are often only found in mature stands (Ram et al. 2020). In all regions, it is the availability of nesting sites rather than tree species that matters.

FIGURE 8.4 Wrynecks will breed in grassy clearings with old trees, surrounded by otherwise dense forest. Pilis Hills, Hungary (GG).

Orchards and vineyards

Some studies have highlighted that orchards and vineyards can be good breeding habitat for Wrynecks as long as there are patches of bare, open ground where they can forage (Coudrain et al. 2010; Weisshaupt et al. 2011). Though non-natural, traditionally managed orchards are ecologically similar to natural wooded-grassland habitats. Indeed, the birdlife of old orchards is known to be much richer than it is in intensively managed ones (Collette 2008; Vogrin 2011; Kajtoch 2017; Chmielewski 2019) and in some areas they are key habitats for Wrynecks (Bautz 1998).

Vineyards, too, often host the species, as ant abundance is often high. However, if they are too tidy and lack trees with suitable cavities for nesting, vineyards are avoided (Assandri et al. 2018).

A study conducted in Valais, south-west Switzerland, focused on the foraging preferences of three insectivorous ground-feeding birds living in such cultivated habitats: Hoopoe *Upupa epops*, Woodlark *Lullula arborea* and Wryneck (Schaub et al. 2010). It was found that when feeding their chicks, all three preferred to forage in mosaics of grass interspersed with bare ground, which orchards and vineyards often provide. However, other studies have not found a strong relationship between Wryneck occurrence and the extent of bare patches (Mermod et al. 2009). Interestingly, in some regions pairs do

FIGURE 8.5 An abandoned orchard that has plenty of trees with holes and hollows, and grassy ground free of herbicides and pesticides. Such places are often ideal breeding habitat for Wrynecks. Komárom-Esztergom County, Hungary (GG).

breed in intensively managed orchards and vineyards. It seems that as long as there is prey, the ground vegetation is not overgrown and there are adjacent nest site possibilities, the species can persist in such places (Zingg et al. 2010).

Urban habitats

It is highly unlikely that the human trend towards urbanisation will abate. This will be unfavourable to most wildlife, yet some species tolerate such change better than others and some can adapt. Although often overlooked by conservationists, urban and semi-urban areas do support some species. Indeed, compared to seed-eaters, insectivorous birds in general can do reasonably well in such secondary habitats (Yosef and Zduniak 2011). As main habitats dwindle, parks and gardens can serve as refuges for Wrynecks as long as there are nesting trees and foraging areas with ants (Reichlin et al. 2009).

FIGURE 8.6 Gardens in small towns, which are maintained in a nature-friendly manner, with untreated grassy areas and old trees left in place, can be havens for Wrynecks when the management of surrounding farmland is intensified. Traisen, Austria (GG).

FIGURE 8.7 The concrete yard of an abandoned state cattle farm may seem unlikely Wryneck habitat. Yet, former industrial sites like this are often occupied as ants can be abundant and adjacent trees provide nesting possibilities. Pest County, Hungary (GG).

Nonetheless, they are not always ideal due to some obvious factors and some which are more subtle. A study of birds of conservation concern in Italy found that urban environments were avoided by most specialist species but some, including the Wryneck, did not necessarily decrease in them, with the exception of town and city centres (Sorace and Gustin 2010). Yet, other studies have shown that Wrynecks will abandon urban areas where conditions seemed, at least to human eyes, suitable. In a long-term study in a park in Rome, Wrynecks eventually disappeared although some other species continued to thrive (Battisti and Dodaro 2016). It was concluded that the reasons were not local but connected to the overall decline of the species. In eastern Europe, former military bases which have reverted to grasslands and abandoned industrial and mining sites with pioneering vegetation have been occupied (Becker and Tolkmitt 2007).

Wintering habitats

In the Mediterranean basin, habitats which are used in the breeding season, such as oak woods, olive groves and orchards, are also frequented in winter. In addition, island and coastal wetlands, reedbeds, maquis/macchia scrub, pine woods, plantations and suburban gardens are all visited (Orsini 1997; Brichetti and Fracasso 2020). A study of inland wintering sites in Extremadura, Spain, found that Wrynecks chose lowland, diverse, irrigated croplands, such as rice fields, which were often close to wetlands, oak woods or groves (van Wijk and Tizón 2016). In the Middle East, vegetated and cultivated areas of land, including scrub, wadis, gardens in settlements and riverine woods are usually occupied. In India they often winter in agricultural areas, even around cotton and soya fields (Kasambe et al. 2014). In South-East Asia, sparse trees and shrubs in open country with farmland, rice fields and fishponds are often used, as well as parks and plantations. Elsewhere in southern Asia, birds can be seen in dry woodlands, bushy grasslands, thickets and even open canopy forest (Winkler et al. 2020). In Africa, most Wrynecks winter south of the Sahara Desert but north of the Guineo-Congolian forests. Some reside in the arid grasslands and acacia savanna of the Sahel (Zwarts et al. 2009), but the bulk occur further south in the open woodlands of the more tropical Sudanian and broad-leaved Guinean forest-savanna zones. They are scarce in the semi-desert grassland and shrubland of the Somali-Masai zone (Zimmerman et al. 1996). In West Africa, they are found in a variety of habitats, including woodland, riparian forest edge, and low thickets in farm bush or

FIGURE 8.8 In Iberia Wrynecks often spend the winter in olive groves. Baeza Jaén, Spain (JMFDF).

FIGURE 8.9 When on migration Wrynecks often occur in arid, treeless terrain such as this semi-desert in Kazakhstan (SA).

forest clearings (Dowsett-Lemaire and Dowsett 2014). In the north of this area, the species tends to favour riparian forest, avoiding more open, dry wooded grasslands.

Habitats on migration

When on migration Wrynecks stop in a varied range of habitats. In the Mediterranean they are often observed in coastal scrub and in both natural and planted pine woods, olive and citrus groves, and on rocky, quite barren islands. In northern Africa and the Middle East birds drop into wadis and oases. In South-East Asia they are seen in mangroves and cultivated areas. Individuals moving through Britain are mostly found on the coast and offshore islands. Some occasionally turn up in gardens (Wernham et al. 2002). As Wrynecks on passage are obviously not breeding, trees are not essential and the main resource needed is food, essentially ants. Hence, birds will often stop in areas which are almost treeless to forage and refuel, such as montane farmland, coastal headlands, beaches, islands and deserts (Winkler et al. 2020).

Chapter 9

Challenges and Conservation

The overall ecology of the Wryneck is poorly known. Most research on this species has been done in western, central and northern Europe, particularly on pairs breeding in nest boxes or as part of general migration studies. Hardly any research has been undertaken in the eastern and Asiatic parts of the species' breeding range and in its wintering quarters. The exceptions are a few ringing studies carried out at the International Birding and Research Centre in Eilat, on the Red Sea in Israel, where migrating birds using the eastern Mediterranean flyway have been caught (Yosef and Markovets 2009; Yosef and Zduniak 2011). Unfortunately, there appears to be no national nor international conservation activities that focus solely on Wrynecks, although locally there are nest box projects (see 'Nest box projects' at the end of this chapter).

Conservation status

The International Union for Conservation of Nature (IUCN) focuses on nature conservation and the sustainable use of natural resources. Although the Wryneck's global population is considered to have decreased, the species is not believed to be vulnerable or seriously threatened as it occurs over an extremely large range. For this reason, in the IUCN's Red List of species, it is categorised as Least Concern (BirdLife 2021). The European Bird Census Council (EBCC 2020) states that the Wryneck underwent a Moderate Decline between 1980 and 2017. Wrynecks are protected by law in all European Union member states and the United Kingdom, but elsewhere in the world the species is often not legally safeguarded, its status varying from country to country.

FIGURE 9.1 An unfortunate Wryneck shot by a poacher when passing through Lebanon (CABS).

Threats

Although Wrynecks are locally common in some regions, their numbers can be unpredictable, and they are sometimes inexplicably absent from apparently suitable habitat where they once were not uncommon. In much of western and central Europe, a long-term decrease in both numbers and range that occurred throughout the twentieth century has been attributed to several factors. In breeding areas, habitat change or total loss, intensification in agricultural practices, inappropriate woodland management and the general industrialisation of the countryside are all cited (BirdLife 2021). Yet, as mentioned in Chapter 7, the demise of Wrynecks as breeding birds in Britain began in the 1800s, before some of these changes commenced, and therefore other factors appear to have been involved. The fact that most studies have been conducted in Europe and have focused on pairs using artificial nesting sites, may have also produced some bias in results and in conclusions on the species' breeding ecology. Both within and away from Europe, climate-related issues and land-use in wintering areas (van Wijk et al. 2013) and on migration routes (Yosef and Zduniak 2011) are also suspected to have played a negative role.

Farming

It is widely accepted that many so-called farmland birds are in decline as a result of agricultural intensification. Although Wrynecks are not farmland birds per se, they often breed in agricultural areas due to their liking for open habitats where there are plenty of ants as well as suitable sites for nesting (Weisshaupt et al. 2011). Farmland accounts for 39% of land-use in Europe and as such can sometimes provide important habitat, especially as so much natural habitat has been degraded or lost entirely. While the species is often considered to be able to live only in traditionally farmed areas, quite intensively managed land is occupied so long as a diverse patchwork containing ants and breeding cavities is present (Mermod et al. 2009). Large tracts of unspoilt terrain are not essential, and breeding success can be high in places that appear suboptimal to humans (Brünner and Rödl 2018). In some regions, pairs may even be dependent on farmland (van Wijk and Tizón 2016). Nevertheless, the species has clearly been adversely affected by the implementation of industrial agricultural methods (Coudrain et al. 2010). The general trend of farmland becoming uniform has diminished its suitability for birds like the Wryneck which thrive in mosaic environments. The removal of trees and hedgerows has meant fewer nesting sites, and the conversion of grassland to cultivation, which effectively destroys ant colonies, has meant less prey. Where grassland has remained, the dense sward that results from the heavy use of fertilisers may be good for cattle, but it is not suitable for terrestrially foraging birds, which need bare ground within grassy habitats (Mermod et al. 2009; Schaub et al. 2010). A study in Switzerland which followed seven radio-tracked Wrynecks in the breeding season found that higher and lusher ground vegetation not only reduced the number of ants but also hindered the birds' ability to access them (Weisshaupt et al. 2011). Fewer grazing animals, as when livestock is taken indoors, can result in vegetation becoming overgrown and as a result the habitat becomes unsuitable for most ants. In parts of Fennoscandia, pastures have sometimes been planted with coniferous trees, ultimately changing the open habitat to a densely shaded forest (Linkola 1978; Ryttman 2003). The indiscriminate use of pesticides applied to cultivated land not only kills invertebrate pests but also exterminates non-target species, including ants. Studies across Europe, including in Finland (Vaisanen 2001), Sweden (Ryttman 2003) and Switzerland (Weisshaupt et al. 2011), have all shown that, when ants become scarce, so do Wrynecks.

Old and new orchards

Across much of Europe, including Britain, the second half of the twentieth century saw cultivation practices in many old orchards intensified, or they were simply cut down and replaced with more profitable enterprises. For example, more than 5,000 ha have been lost in Slovenia since 1990 (Vogrin 2011), in north-west Germany 75% were lost between 1979 and 2009 (Forejt and Syrbe 2019) and in Normandy, France, around 600,000 ha in 1950 had shrunk to just 146,000 ha by 2000 (Collette 2008). Newly planted orchards tend to be uniform, intensively managed monocultures with few nesting possibilities and heavy pesticide use. Pre-industrial, cultural landscapes often included orchards, which were probably important for providing many of the resources on which Wrynecks depend (Mikusiński and Angelstam 1997), and the trend away from traditional management has been detrimental (Gedeon et al. 2014).

FIGURE 9.2 Traditionally managed orchards with open grasslands are ideal Wryneck habitat. Once they are intensified the birds begin to disappear. Eschenau, Austria (TH).

Woodland management

Wrynecks have declined in wooded habitats, too. When deciduous and mixed forests or woodlands are replaced by denser stands of conifers, and grassland openings become afforested when grazing animals are removed, these birds disappear. The intensification of forestry management also reduces nesting sites. In conifer plantations, trees are felled before they reach an age when cavities (natural and woodpecker-created) appear in them. Indeed, some studies have suggested that the availability of suitable nest sites is the single most important resource for Wrynecks (Zingg et al. 2010). The understorey in forests often becomes overgrown, too, as when thinning or selective felling open the canopy, letting in more sunlight. Yet, this species can benefit for a time when forests are clear-cut as the open areas created often host high densities of ants (Kervyn and Xhardez 2006; Bijlsma 2014). Salvage logging is a procedure in which damaged timber is removed soon after fire or high winds to prevent economic losses due to the activity of wood-boring insects and fungi. A Swedish study of the impacts of such felling on birds concluded that,

FIGURE 9.3 Trees that are hollow or contain holes, such as this former woodpecker cavity, should not be felled. Appallingly, this tree was cut in the nesting season. April 2015, Zemplén Hills, Hungary (GG).

although it could be beneficial for species that typically occur in open wooded landscapes, it was harmful to overall forest-species richness and abundance. Of the six woodpecker species present in the study area, only Wryneck, as an open-landscape species, favoured salvaged plots (Zmihorski et al. 2019). In addition, Wrynecks can sometimes benefit from afforestation. For example, in Denmark they have found new breeding habitat where conifers have been planted on dunes (Andersen 2018).

Weather and climate

Changes in climate have also been suggested as having adversely influenced Wryneck populations (Vogel 1997). In the breeding season, feeding behaviour, nestling growth and, ultimately, their survival can all be negatively affected by prolonged periods of heavy rain and both high and low temperatures. Unfavourable weather forces ant prey to move deeper underground, and thus foraging becomes more difficult (Geiser et al. 2008). This may explain why populations in northern and north-western Europe, where a wet and cold maritime climate prevails, have been more prone to population and range contractions. On the other hand, it has been suggested that rising average temperatures may result in the Wryneck, among other bird species, expanding its range and recolonising areas from which it has disappeared (Price 2017).

Wintering areas

Most of the European population winters in Africa, but, although this certainly involves millions of birds, the majority seem to vanish once there (Zwarts et al. 2009). Indeed, in most wintering areas numbers are unknown, mainly owing to the low detectability of the species in that season. The paucity of ringing recoveries from Africa also means that only inferences about the locations of the main wintering sites can be made (Reichlin et al. 2009). Consequently, even less is known about the ecology of Wrynecks in their winter quarters than in their breeding areas. In Africa, climatic and land-use changes may well be adversely impacting the species (Sanderson et al. 2006). The decline in Europe might also be linked to precipitation patterns in Africa, specifically to a prolonged drought that occurred in the Sahel from the 1970s to the early 1990s.

A detailed study in that region found a positive correlation between the population sizes and survival rates and precipitation of several species of

FIGURE 9.4 Richard-Toll, Northern Senegal, near the Senegal River and the Mauritanian border, January 2007 (NiB). The land here is parched, degraded, overgrazed by goats and the acacia trees heavily hacked for wood and fodder. Yet, for the moment, Wrynecks continue to winter here.

European breeding birds that migrate long distances, including the Wryneck (Zwarts et al. 2009). It was noted that birds which winter in the wooded savannas were especially afflicted, and habitat degradation and loss were suspected as being two of the causes of the Wryneck's decline there. After decades of drought, rainfall began to increase in the 1990s and it was seen that as it did numbers of several migratory birds also began to increase. Even so, ongoing deforestation in Africa probably does not play a significant role as Wrynecks are not overly arboreal when not nesting. In addition, as already mentioned (see Chapter 7), some European populations of the species do not winter in Africa but in the Mediterranean basin (van Wijk and Tizón 2016). Clearly, those that remain in Europe cannot be affected by local conditions in Africa and it is likely therefore that a combination of factors, in both breeding and wintering areas, and along migration routes, has been detrimental.

Hunting and trapping

Birds face some daunting challenges when on migration. Besides the natural hazards that all migratory species encounter, human activities often make their journeys even more perilous. Although Wrynecks are generally not specifically targeted, they do suffer from being hunted and trapped, particularly in southern Europe, the Middle East, North Africa and South-East Asia. For example, most ringing recoveries involving dead birds in Italy are believed to be the result of 'human activities' (Spina and Volponi 2008). Large-scale netting sites in Cyprus, Lebanon and Egypt are a huge threat with countless numbers of birds trapped and killed in these three countries alone each year. In Cyprus lime-sticks are also used.

During autumn migration across Egypt's Sinai coast, long lines of trammel nets are set to catch Common Quails *Coturnix coturnix*. These nets ensnare large numbers of that target species, but also illegally capture many others,

FIGURE 9.5 A Wryneck caught in an illegal net in Cyprus (CABS).

FIGURE 9.6 A Wryneck trapped in the gluey-slime of a lime-stick in Cyprus (CABS).

including Wrynecks, which are regarded as useless and so destroyed. Surveys conducted from 2008 to 2012 along the coast of North Sinai found that, in addition to quails, a further 54 bird species were caught in such nets. The numbers are alarming, with rates of capture from 2008 to 2012 estimated at 2 million quails and 0.5 million birds of other species slaughtered annually during the 45 days of peak migration. It is believed that more than 8500 Wrynecks per year are caught in nets in North Sinai alone (Eason et al. 2015). Sadly, in that region and others, hunting and trapping (legal and illegal) are likely to continue to be a major problem contributing to population declines of migratory birds until laws are implemented and then enforced.

Action and solutions

The underlying causes of the Wryneck's decline are not fully understood and therefore it is difficult to propose specific conservation measures. Furthermore, as this is essentially a migratory species, any conservation management would not be straightforward. Unlike the situation with sedentary birds, potential problems in stopover and/or wintering sites (where it is likely that most mortality occurs) also need to be considered. The Wryneck is one of several Afro-Palearctic migratory species that have shown a trend of continuously decreasing numbers. In common with other inter-continental migratory birds, it has been negatively affected more than either short-distance migrants or resident species (Sanderson et al. 2006). This tendency is thought to be related largely, although not entirely, to problems in its African wintering habitats. Clearly action is required outside Europe to stem the decline, although more research is needed to assess where in the world the main causes actually lie (Schaub et al. 2012).

Nevertheless, at a local level, various do-able conservation activities can be undertaken. On existing reserves, management measures should not neglect the maintaining of semi-open landscapes with patches of sparsely vegetated ground to ensure good foraging conditions, not only for Wrynecks but also for other terrestrial feeders (Schaub et al. 2010). On farmland, even in intensively worked areas, conditions similar to those just mentioned for reserves can be established and pesticide and fertiliser use reduced or, better still, terminated. Indeed, across the European Union and the United Kingdom agri-environment schemes have the potential to positively change land management by paying farmers to implement measures to improve habitats for a range of birds including the Wryneck. Wherever possible, existing suitable habitats in farmed landscapes should be protected. In vineyards and orchards, traditional, low-intensity practices should be encouraged to maintain ant-rich habitats. Large limbs on fruit trees should not be lopped nor old trunks removed, as these are the places where cavities form. In woodlands, too, trees with hollows and woodpecker holes should be retained. Woodlands can be multi-structured to include edges and open areas such as grassy rides and glades which host more insect life, including ants. In semi-urban habitats, such as parks, any dead or dying trees, which may contain cavities, should likewise be left in place. Habitat restoration and creation can also provide additional habitats for Wrynecks and a host of other species.

Nest box projects

In woodlands, particularly plantations, where there are few or no old trees with natural cavities or old woodpecker holes, nest boxes can be placed out. Nest box projects are a popular and relatively simple and inexpensive way of assisting secondary cavity-using birds. In managed habitats where cavities are lacking but adequate foraging is available, boxes can mean the difference between Wrynecks breeding and not breeding. Pairs will sometimes even settle in new areas after boxes intended for them have been provided (Zingg et al. 2010). For more on nest boxes see Chapter 13.

FIGURE 9.7 Nest box for Wryneck placed by the author on the edge of a conifer plantation. Such woodland monocultures seldom have cavities of any kind but often have suitable foraging areas. Pest County, Hungary (GG).

Chapter 10

Behaviour

Wrynecks are diurnal. They are most active in the morning hours; in the middle of hot days they often perch in cover, typically resting with their head and tail lowered. If approached at such times, individuals usually react by remaining still, apparently relying on camouflage, although some will be flushed and fly to another nearby area of cover.

FIGURE 10.1 What's in a name? A Wryneck with its neck awry. September 2018, Bourgas, Bulgaria (GG).

Sociability

The Wryneck is not an overly sociable bird. Of course, members of a mated pair associate with one another during courtship and when raising their brood and form a strong bond at that time, sometimes even spending time together in the cavity in the pre-incubation period. This relationship, however, probably lasts for only one season. When on migration, several individuals may be observed at stopover sites, but this is almost always due to weather-induced 'falls' or to an abundant supply of food in one area, rather than to any social or flocking instinct. After fledging, the young usually stay together for just a few days before separating, although parents will continue to feed them individually in the vicinity of the nest for up to three weeks (pers. obs.).

Roosting

During the breeding season one adult will roost overnight in the nest cavity, incubating the eggs and later brooding the chicks. The adult squats on the eggs or small nestlings, but later clings to the cavity wall (Menzel 1968). Cavities in the vicinity that are not used for breeding are utilised as roosting sites. Occasionally, some recently fledged young, or a parent, may return to the former nest to roost. Where Wrynecks roost when stopping on passage largely remains a mystery. Presumably, they perch in the open, in trees and bushes, but sometimes they can be inventive. A report from Jersey, Channel Islands, describes a Wryneck clinging to the top of a single reed in a reedbed among wagtails, swallows and other roosting migrants (Young et al. 1993).

Comfort behaviour

Like all birds, this species indulges in comfort behaviour, a term which describes activities related to body and plumage care, and which is thought to enhance physical wellbeing. Preening, scratching, stretching, basking, dusting, bathing and anting are all forms of this behaviour.

Preening: This is probably the most commonly observed form of comfort behaviour for all birds. It is a cleansing process, an essential part of plumage maintenance whereby oils are applied and dirt and ectoparasites removed. Preening typically involves feather-shaking and feather-ruffling, and the precise drawing of individual feathers though the mandibles. Wrynecks do not allopreen, that is, individuals do not preen one another.

FIGURE 10.2 A parent takes a break from incubating the eggs and indulges in a bout of preening and sunning at the entrance to its nest. May 2021, Ipolydamásd, Hungary (GG).

Sunning: Basking in the sun normally takes place on the ground. This activity is thought to enhance feather health as birds often begin to preen during or immediately after sunning. It probably also helps in the production of vitamin D, which is a key component in preening oil, and discourages parasites such as mites and ticks.

Bathing: Although Wrynecks will bathe in water (Glutz and Bauer 1994), this behaviour is rarely observed as these birds are especially vulnerable and hence cautious at such times. Yet it may be more common than presumed as Wrynecks are habitually on the ground foraging for prey and so probably often come upon shallow pools and puddles.

Dusting: For the same reasons as for water bathing, dust bathing is seldom observed. It is unclear whether Wrynecks use regular sites or whether they dust opportunistically. Dusting may be more common than reported as the birds spend so much time on the ground and often select places with a high proportion of exposed dry soil (Weisshaupt et al. 2011). Dusting probably helps to eliminate parasites and therefore, as with sunning and bathing, is most likely undertaken as a means of maintaining plumage health.

Anting: This is a specific form of behaviour in which birds permit, or encourage, ants to squirt onto their feathers the chemicals they release when disturbed. There are two forms of anting: passive and active. In passive anting a bird simply squats down among the ants and allows them to swarm over its body, the bird merely spreading its wings and tail and not attempting to pick up the insects; it may also wriggle around, perhaps to provoke the ants to swarm and eject their chemicals. Active anting involves the bird applying the insects directly to its body, preening with them in its bill. Elements of both passive and active forms are sometimes combined. The precise function of this behaviour is disputed (Morozov 2015). The most common explanation is that it is a type of comfort behaviour, the formic acid that the insects eject aiding feather care and countering parasites. Indeed, birds often preen during or immediately after anting, and the chemicals may reduce skin irritation during the moult of the feathers. It has also been suggested that it is a form of food preparation in which ants are urged to discharge their acid so that they thus become more palatable (Judson and Bennett 1992). Wrynecks practising anting is seldom observed (Stone 1954; King and Speight 1974).

Displays

Although Wrynecks differ from their woodpecker relatives in many ways, their displays are comparable. Both courtship and antagonistic encounters involve rhythmic swaying and rotating of the head, neck-stretching, bill-pointing, ruffling of the crown feathers, tail-spreading, drooping and spreading of the wings, and aerial chases (Menzel 1968; Löhrl 1978). Slow head-swaying, which is often done at incubation change-overs, is probably a form of appeasement (Ruge et al. 1988). For specific details of courtship displays see Chapter 12.

FIGURE 10.3 Wrynecks will ruffle and raise their crown feathers in various situations, such as when excited, alarmed or agitated. September 2018, Bourgas, Bulgaria (GG).

Aggression and defence

Wrynecks are territorial, being especially possessive of nesting cavities. They are generally hostile to other cavity nesters, sometimes even confronting larger species. When faced by a conspecific challenger, a breeding Wryneck will typically respond by direct bill-pointing, neck-stretching, raising of its head and erecting its crown feathers. Actual fights, with bill or body contact, are uncommon. When faced with a predator outside the nest, whether avian or mammalian, there is much individual variation in response. Some birds will purposefully sway their head from side to side, or when on the ground raise their crown feathers and fan out their tail. Others will freeze, squatting along a branch, not unlike a nightjar, seemingly relying on crypsis to escape detection (Glutz and Bauer 1984). They may also fly at and around the threat,

FIGURE 10.4 Reactions to apparent threats include hiding in cover or, like this bird, behind a branch. April 2021, Gerecse Hills, Hungary (GG).

including humans, rapidly fluttering their wings which make a whirring sound, occasionally directly mobbing it. Some individuals simply take flight (Gorman 2021). Nestlings usually pause their begging calls when alerted to an approaching predator.

Mimicry

The most renowned behaviour of the Wryneck is probably its 'snake imitation' in which, when threatened or captured, birds perform stereotyped writhing and twisting head and neck movements and/or utter hissing sounds (Steinfatt 1941). When they perceive danger while in the nest cavity, some individuals open their bill, slowly stretch out their neck, point their head towards the threat and then, when fully extended, suddenly retract it (Ruge 1971). Such movements and sounds are regarded as classic examples of behavioural mimicry. This is simply defined as when a harmless species attempts to deter a potential predator by mimicking a harmful one (Stoddard 2012).

Two forms of this mimicry are generally described, one visual and one vocal. Wrynecks are considered to use both: visual when they turn their head,

FIGURE 10.5 In response to being caught this individual is in full-on writhing mode, bending its head right back. September 2018, Bourgas, Bulgaria (GG).

and vocal when they utter hissing sounds. Additionally, it has been suggested that they specifically imitate vipers (*Vipera*) as the blackish line on the bird's back resembles the dorsal zigzag line of these snakes, a pattern which is believed to signal to predators to avoid them (Brejcha 2019).

Some exaggeration has evolved around this intriguing behaviour, particularly in anecdotal reports but also in the literature. It is not unique to this species (as mentioned in Chapter 5, Red-throated Wrynecks perform similar displays). Globally, many other birds will also twist and rotate their head and make snake-like sounds if disturbed or confronted by a potential predator, particularly cavity or burrow nesters when incubating or brooding chicks. Some true woodpeckers and cavity- and burrow-nesting owls also produce stereotyped sounds that resemble snake hisses when a potential predator

FIGURE 10.6 A Wryneck at its nest hole in an apple tree. The bold dark dorsal stripe has been compared to the zigzag line on the back of vipers. May 2021, Ipolydamásd, Hungary (GG).

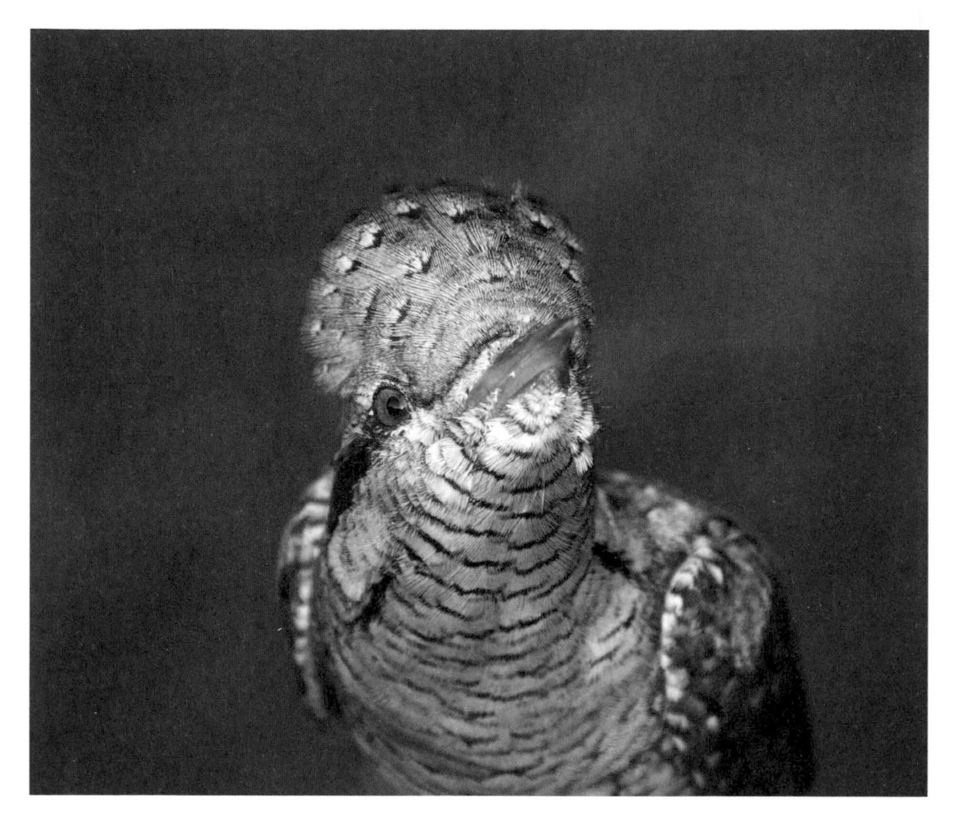

FIGURE 10.7 When alarmed Wrynecks often spread and raise their crown feathers. May 2012, Mohelno, Czech Republic (TG).

approaches. The Common Kingfisher *Alcedo atthis*, a tunnel-nester, will turn its neck when handled. Both adults and chicks of titmice and chickadees (Paridae) in North America produce hissing sounds and lunging movements (Sibley 1955). In Europe, these behaviours have been studied in detail in the Great Tit *Parus major* (Krams et al. 2014; Møller et al. 2021) and the Blue Tit *Cyanistes caeruleus* (Dutour et al. 2019).

Furthermore, not all Wrynecks react in such ways when threatened, and those that do so differ greatly in their responses (Gorman 2021). Stimulus thresholds vary considerably. Some may flap their wings, some only droop them, while fanning their tail and raising their crown feathers. The bill may or may not be opened and calls may or may not be given during these movements. Besides, not all the sounds made can be reasonably described as snake-hissing: some have been compared to those of a frightened cat and even to a fizzy drink bottle being opened (Glutz and Bauer 1994). When discussing vocalisations Ruge (1988) does not mention hissing but does state that occasionally birds make sounds when they feel threatened in their nest.

Some individuals are passive, allowing themselves to be quietly removed from the nest, or they withdraw into the bottom of the chamber and squat on their clutch or brood. Large nestlings, too, tend to remain silent and huddle together when inspected in nest boxes. Other individuals, including fully feathered nestlings, panic and try to evade capture (Gorman 2004). Much seems to depend upon what the intruder is doing and for how long. The stage in the breeding cycle is probably another factor. When captured and handled by ringers, most Wrynecks twist and rotate, but some do so only briefly (unless provoked) or merely sway gently from side to side. Occasionally a bird does not 'perform' at all. When alleged snake mimicry does take place, it is invariably by a bird which cannot escape, owing to being held or cornered. It has been suggested that this behaviour occurs in cavity-nesting birds because snakes are common inhabitants of such sites, and because cavity nesters have few

FIGURE 10.8 A Wryneck with its tongue extended out of its bill. This is not an attempt to imitate a snake, but quite common behaviour done in various situations, including when birds are not threatened. June 2012, Fejér County, Hungary (AK).

possibilities to exit from the enclosed space and so resort to deterring potential predators (Krams et al. 2014). Given the chance, Wrynecks invariably try to flee (Gorman 2021).

As they grow, nestlings begin to flick their tongue in and out, but this 'tongue-darting' is not snake mimicry as it is done routinely by both young and adults in situations when they are clearly not threatened. In fact, young woodpeckers of many species 'try-out' their tongues in this way. Nestlings make various rasping, zizzing and sizzling calls when begging to be fed and sometimes do so when a human inspects a brood in a nest box, probably responding instinctively to the presumed arrival of food. They also make sharp chirps which can develop into whirrs (Glutz and Bauer 1994). With a little imagination some of these calls could be interpreted as imitations of the sounds of a snake.

Sometimes chicks open and close their bill without vocalising, which may be an instinctive reflex related to the possibility of a feeding visit by their parents. Outside the nest, for example, when being ringed, adults hardly ever give hissing sounds, although there are reports of them (Steinfatt 1941). If they call at all, they are more likely to emit loud, shrill alarm notes. In response to threats when outside the nest but not in the hand, Wrynecks do not use perceived mimicry as a means of defence.

Ultimately, no conclusive evidence has been presented to prove that any of this behaviour is truly snake-mimicry functioning as anti-predator display. It might also be argued that loud hissing is simply a disconcerting, startling sound aimed to distract, and that mimicry is not involved. In addition, nestlings utter subdued grating and hissing sounds when alone in the nest, between feeding visits and when not threatened (pers. obs.). Mimicry, especially visual, does appear to be an important anti-predator response in evolutionary ethology but further research on this fascinating subject is required to support it in the Wryneck, especially as regards hissing calls, which appear to be rare (Gorman 2021).

Feigning death

When handled, some individuals use another strategy to defend themselves. They lie flat, hanging limp, often with their eyes closed, seemingly pretending to be dead. This, too, might be regarded as a form of mimicry as some snakes also use this defensive strategy. Wrynecks that feign death in this way are perhaps also trusting in their cryptic plumage as camouflage.

FIGURE 10.9 When handled, most individuals, like this one about to be ringed, are passive and some seem to 'play possum'. September 2006, Hartberg, Styria, Austria (TH).

Displacement activity

Wrynecks do not always react in dramatic ways to perceived danger. Like many animals, when faced with a threat they may engage in acts that seem to have no obvious connection to the situation in which they find themselves. This is termed 'displacement behaviour'. For example, a disturbed Wryneck may suddenly begin to preen, forage or gently peck with the bill on branches near its nest (Gorman 2004). Individuals often seem agitated when they indulge in these mock behaviours. Sometimes they react by totally ignoring the threat rather than engaging with it, apparently feigning disinterest. Wrynecks do not indulge in paratrepsis, which is a more extreme distraction behaviour rather than displacement response.

Chapter 11

Flight, Movements and Migration

The flight action of woodpeckers is often described as 'bounding', birds flying with a series of fast wingbeats followed by glides on closed wings, dropping and rising as they progress. Yet, this pattern is not characteristic of many woodpeckers (Gorman 2014). Despite some descriptions stating that Wrynecks fly in this manner, they typically do not.

On brief flights, such as when changing perches, they tend to fly low, straight and rapidly. When moving between the nest and foraging areas, the length of flight is also often short, sometimes just a few hundred metres, but can increase as the breeding season progresses. This is likely to be due to the depletion of the ant colonies nearest the nest (Bijlsma 2014).

FIGURES 11.1 and 11.2 An adult flying to and from the nest tree. Novo Yankovo, Bulgaria (TA).

FIGURE 11.3 An adult drops low and away from its nest box. June 2021, Balatonmáriafürdő, Hungary (IB).

Migratory picids

Many woodpeckers globally undergo local dispersive movements and some periodically move longer distances, usually in response to harsh weather and/or depleted food resources. In uplands some make seasonal elevational movements. Few woodpeckers, however, are truly migratory, the majority being resident and largely sedentary. Genuinely migratory species occur mostly in the Americas, for example, Yellow-belled Sapsucker *Sphyrapicus varius*, some populations of Lewis's Woodpecker *Melanerpes lewis* and Northern Flicker *Colaptes auratus*. Those which breed in Canada and the northern United States move south in autumn to the southern United States, Central America and in the case of the sapsucker to Caribbean islands before returning north in spring. In Asia, the *subrufinus* subspecies of Rufous-bellied Woodpecker *Dendrocopos hyperythrus*, which breeds in Manchuria in China and south-east Russia, migrates to winter in southern China and Tonkin in Vietnam (Gorman 2014). The only other Old World woodpecker that is migratory is the Wryneck.

Distance and time

Some populations of Wrynecks are long-distance migrants, some short-distance, whilst others are resident. The long-distance migrants include those that breed in northern Europe and winter in sub-Saharan Africa, and Central Asian populations that winter in India and South-East Asia. The total journey from northern Europe to African wintering areas could be up to 6,000 km. Short-distance migrants include central European breeders who fly to Iberia. Tracking devices have revealed that some Swiss and German birds move about 1,500 km to Spain (van Wijk et al. 2013).

Ringing recoveries can also reveal something about the distances Wrynecks migrate and the time their journeys take. For example, most autumn recoveries from Swedish ringed individuals are from sites over 1,000 km away and are distributed across southern Europe from Portugal to Italy, and also in North Africa. The nearest recovery was from southern Germany, some 725 km away (Hedenström and Lindström 1990). To date there is no recovery data from beyond the Mediterranean for Wrynecks ringed in Hungary. Indeed, most Hungarian recoveries are very local, often within 5 km (Török 2009). Still,

FIGURE 11.4 Ringing studies have helped shed some light on Wryneck migration strategies. September 2006, Hartberg, Styria, Austria (TH).

these records are only indications and do not demonstrate how far the birds concerned actually travelled or where they wintered.

Most Wrynecks probably move at night so there is limited information on how long it takes them to reach their destinations, although individuals trapped by day have given hints. For instance, a bird caught at the Azraq ringing station, Jordan, on 18 April 1965, was re-trapped there five days later, and then found dead on 9 May at Kozelets, Ukraine, around 2,100 km to the north-west (Peal 1973). Another bird recovered (after being shot) in Beirut, Lebanon, in October 1988 had been ringed eight days earlier in Central Anatolia, Turkey, a distance of 600 km (van den Berk 1990). Another ringed in southern Finland was caught in Hungary two months later after having travelled an estimated 1651 km (Török 2009).

Spring timing

At Lake Chad in northern Nigeria, Wrynecks are recorded on passage mainly in February and March (Gustafsson et al. 2021) but as late as early May (Dowsett 1968). Most move through Mauritania from March to mid-May (Isenmann et al. 2010). They pass through Algeria from late March to early May (Short 1988a; Isenmann and Moali 2000). In the east, peak numbers move through Ethiopia and Eritrea in March (Ash and Atkins 2009). The last birds leave Oman in late April (Eriksen and Porter 2018) and pass through the United Arab Emirates mainly in March and April (Richardson 1990). At Eilat most are trapped in mid-March (Yosef and Markovets 2009; Yosef and Zduniak 2011). The majority pass through the rest of Israel mainly between the third week of March and the third week of April (Shirihai et al. 1996). Just to the north, they pass through Lebanon from early March to mid-May (Ramadan-Jaradi et al. 2020). Ringing records show that Wrynecks move through the Azraq Wetland Reserve in eastern Jordan from mid-March to late April (Stêpniewska et al. 2011). Most are seen in Iran from late March to late April (Tohidifar 2008). In the Far East, birds pass through the Korean Peninsula in April and May (Winkler et al. 1995), and in Japan they occur from May in northern Honshu and Hokkaido (Chikara 2019).

The appearance of Wrynecks in European breeding areas is staggered, being anytime between March and May. In the eastern Mediterranean birds move through Cyprus from February to May and peak from late March to mid-April (Richardson and Porter 2020). Further east, the first appear in Armenia in mid-April (Adamian and Klem 1999). In central and western Europe, the majority arrive in early April. Most move through Malta from the end of March to late April. In Italy spring movements on the mainland occur between February and early June, peaking in numbers in mid-April (Brichetti and Fracasso 2020). On Vivara Island, off Naples, the first birds arrive in March (Scebba and Lövei 1985). The first birds are seen in Hungary in mid-March with numbers peaking in mid-April, although most are ringed in early June owing to the dispersal of recently fledged local birds (Török 2009). Spring passage occurs on Helgoland and Schleswig-Holstein in northern Germany, from mid-April to late May (Busche 2004). In Britain, the first birds usually occur in late March, the latest in late May (Fraser and Rogers 2006). A study of several migratory species in Estonia, including the Wryneck, concluded that weather, route taken and distance were important factors that determined spring arrival times, but most appeared in the first week of May (Palm et al. 2009). In north-west Russia, the first appear in late March and numbers peak in late April into May (Iovchenko and Kovalev 2016). All in all, most Wrynecks have arrived in northern European breeding areas by mid-May (Winkler et al. 2020).

Changes in arrival times

Average spring temperatures are increasing in Europe and climate change is considered the main reason. It's believed that this warming has resulted in some migratory birds arriving earlier in breeding areas than they did previously. Ringing records from Schleswig-Holstein, for example, show that the mean spring arrival date (23 April) of Wrynecks is one week earlier than it was 60 years ago (Busche 2004). Yet, early arrival dates are not universal. A study which examined changes in the spring arrival dates of 36 bird species in south-east Hungary and western Romania over a hundred-year period, found that the average spring arrival dates of most medium- and short-distance migrants were earlier than they were in the late nineteenth and early twentieth centuries, but for most long-distance migrants there had been no significant changes. Interestingly, spring arrival dates for Wrynecks have been found to be later in recent decades (Bozó and Csörgő 2020). The inherent migratory instincts that long-distance migrants possess may mean that birds like the Wryneck do not react as quickly to changes in the climate of breeding areas as short-distance migrants do. Indeed, long-distance migrants can only respond to environmental cues in their wintering grounds and rely on finding sufficient food prior to making their journeys (Vickery et al. 2014). There are theories that drier winters in sub-Saharan Africa reducing the availability of food may delay their return but, ultimately, it is difficult to draw general conclusions about why birds in some areas are arriving later and in others earlier than they did in the past until more studies have been done.

Autumn timing

The first Wrynecks leave Europe in summer, sometimes as early as July (Winkler et al. 2020). Autumn passage at Lake Ladoga, north-west Russia, is from late July to early October with the peak in August (Iovchenko and Kovalev 2016). In Hungary, numbers peak in the second half of August into September, the last birds moving through in early November (Török 2009). Passage through Helgoland and Schleswig-Holstein, northern Germany, is between mid-August and early October (Busche 2004). In Italy, post-breeding movements begin in late July, with peak numbers from mid-August into September (Brichetti and Fracasso 2020). The majority of Wrynecks from breeding populations elsewhere in Europe are trapped in Italian ringing camps in August and early September, with numbers falling in October and the latest birds passing through in November (Spina and Volponi 2008). In southern Italy birds from further north arrive from late September until late November (Scebba and Lövei 1985). The first appear in Malta in late August with most ringed in September (though only around 50 annually). In the eastern Mediterranean, Wrynecks pass through Cyprus from August to November with a peak in mid-September (Richardson and Porter 2020). In Britain, though numbers are now small, autumn passage begins in early August, peaks in late September, and typically ends in October with some stragglers in November (Wernham et al. 2002).

Further east, Wrynecks move through Central Anatolia, Turkey, from late September into October (van den Berk 1990). In Armenia, the last are usually seen in late September (Adamian and Klem 1999) with a peak in the Caucasus in mid-August (Winkler et al. 2020). In Lebanon they are reported from mid-August to early November (Ramadan-Jaradi et al. 2020), and in Israel most are ringed at Eilat in early September (Yosef and Zduniak 2011). In Iran birds move through from late August to October, the latest recorded date being 30 October (Tohidifar 2008). Further south in Arabia, they pass through the United Arab Emirates in September and October (Richardson 1990), and the first arrive in Oman in late August with the peak in September (Eriksen and Porter 2018). Most move through Eritrea and Ethiopia in October (Ash and Atkins 2009). Those that winter south of the Sahara pass through North Africa from mid-September to early November (Short 1988a) – most are seen in Mauritania from early September (Isenmann et al. 2010). At Lake Chad in northern Nigeria, they are mostly recorded from September to November (Gustafsson et al. 2021). But the first birds can arrive in their wintering quarters from late August (Winkler et al. 2020). In the Far East, most Wrynecks pass through the Korean Peninsula and leave the north of Japan in September (Chikara 2019).

Routes

Wrynecks that breed in Europe are believed to use two main migration routes. A study that assessed ringing recoveries from 1914 to 2005 from across the continent found that autumn routes showed a similar pattern to that documented for other European breeding birds that are long-distance migrants. Wrynecks that breed in western and north-western Europe are thought to winter in West Africa, while most from eastern and north-eastern Europe head to East Africa (Reichlin et al. 2009). The same study also concluded that the longitude (west to east) of the ringing site correlated with gradual variation in migration routes from the south-west to the south-east. Those that breed in the north-west and western Scandinavia follow the Atlantic coast, some passing through Britain, to Iberia and then on to sub-Saharan West Africa. Those from the north-east, such as the Baltic states and Finland, and central-eastern Europe, move through Italy and the Balkans. Hungarian ringing records reveal that birds that breed or pass through the country use this flyway (Török 2009).

Recoveries in Italy show that most birds passing through the country originate from the Baltic states, Scandinavia and particularly Finland, with fewer from north-east France, Germany, Poland, Belarus, Hungary and the Czech Republic (Spina and Volponi 2008). Recoveries in Germany, on the other hand, suggest that many migrate along a north-east/south-west axis with Fennoscandian birds passing through in both spring and autumn. German breeders mostly head towards France and Spain in autumn although there are recoveries from Italy and Holland. Indeed, it is thought that many central European Wrynecks winter in Iberia and north-west Africa and that the same routes are used in both spring and autumn (Bairlein et al. 2014). Yet the presumed flyways are not clear-cut. For instance, birds ringed in Italy have been recovered over a wide area across southern Europe from Spain in the west to Bulgaria in the east (Spina and Volponi 2008). Clearly, Wrynecks cross the Mediterranean over a broad front, but probably use routes that avoid crossing the widest stretches of sea (Wernham et al. 2002). They may island-hop, using Malta, Sicily, Sardinia, the Balearics and the like, but then again,

FIGURE 11.5 The rocky Boguty Valley, Kazakhstan. Wrynecks do not breed here but occur on passage. The origins of these birds remain a mystery. May 2018 (SA).

at the Strait of Gibraltar Bird Observatory (close to where the Mediterranean is at its narrowest), surprisingly few Wrynecks are caught.

A third route, in the very east of the Mediterranean, continues through Turkey, the Middle East and Arabia (Winkler et al. 2020). This flyway does not seem to involve birds from the European populations mentioned above. This is presumed as there are no recoveries of birds ringed in Europe at Eilat in Israel (an important stop on this route) and no birds ringed at Eilat reported in Europe. It's therefore assumed that Wrynecks passing that way come from the eastern extremes of Europe, from Russia and/or from Central Asia, and bypass the Mediterranean Sea entirely (Yosef and Zduniak 2011). The routes taken by populations that breed in Siberia and Asia are less known than those of European populations. Indeed, almost nothing is known about Wryneck movements in the countries of Central Asia such as Turkmenistan, Uzbekistan, Kyrgyzstan and Tajikistan.

Loop migration

Many migratory birds use the same routes in spring, when heading to their breeding grounds, that they do in autumn when moving to their wintering areas. Such fidelity has, for instance, been observed in Wrynecks crossing Spain's Balearic Islands (Garcías 2010). Yet some birds use different routes in these two seasons, shifting between entirely separate flyways. This is known as 'loop migration'. Reichlin et al. (2009) studied seasonal distributions of ringing recoveries and migration patterns of Wrynecks across Europe and found that in autumn most records originated from the western Mediterranean while in spring they came from Sicily and the Greek Islands. In a more localised example, the number of Wrynecks trapped for ringing in the German Bight of the North Sea coast seems to demonstrate that more move through Heligoland in autumn (August–October) than through the Wadden Islands and mainland Schleswig-Holstein. In spring (April–May) the situation is reversed (Busche 2004).

Overshooting

Around 200 individual Wrynecks each year are thought to pass through Britain. Weather conditions significantly influence their occurrence. In spring they arrive with south-easterly winds from across the North Sea; and in autumn easterly or north-easterly winds associated with a high-pressure system over Scandinavia, usually result in arrivals. Single birds are generally observed, which are thought to belong to breeding populations to the north-east in Fennoscandia. Indeed, of just five ringed in Britain and recovered abroad, two were found in Sweden and one in Finland; the others were one recovered to the east in Germany and one in the south in France (Robinson et al. 2020). Recoveries in Britain of young birds ringed in Norway and adults from Sweden also support this notion. In Scotland most are seen or trapped on Shetland, Fair Isle, Orkney, the Isle of May and the east coast of the mainland (Forrester and Andrews 2007). In England they are also found chiefly along the east coast, particularly on headlands. Most occur from the end of August to October, often well dispersed. It is believed that spring birds have overshot as they were moving northwards through continental Europe. Those that turn up in autumn are also thought to be of Fennoscandian origin, particularly when they arrive in 'falls' with other species that breed in that region such as Icterine Warbler *Hippolais icterina* (Forrester and Andrews 2007).

Elsewhere, individuals which occasionally turn up in the Canary Islands in autumn, mainly in August and September (Martín and Lorenzo 2001), have also probably overshot when en route to West African wintering areas.

Leap-frog migration

Some migratory birds employ a strategy where populations from the north of their range fly south and pass over areas where other populations winter. For example, it is thought that Wrynecks wintering in southern Italy, Sardinia and Sicily are residents (subspecies *tschusii*) while birds from further north in Europe (nominate subspecies *torquilla*) fly over and beyond them to winter further south, possibly in sub-Saharan Africa (Brichetti and Fracasso 2020). The latter strategy is known as 'leap-frog' migration, a concept supported by ringing data, in particular measurements taken of wing lengths and body weights of different populations. Northern-breeding Wrynecks trapped when passing through southern Europe in spring are generally longer-winged and weigh less, sometimes significantly after a long flight, than those that have wintered in the region (Scebba and Lövei 1985; van Wijk et al. 2013). However, although the wings of birds caught in the south of France were relatively short, suggesting *tschusii*, researchers there considered it unlikely that they originated from Corsica or Italy where those populations occur (Orsini 1997). In Malta measurements of birds caught for ringing suggest that both *torquilla* and *tschusii* pass through, with *torquilla* more common. Obviously, leap-frogging means that northern birds fly much greater distances than those which stay in the Mediterranean. Why some Wrynecks are long-distance migrants, others short-distance migrants, and some non-migratory, is not understood.

Stopovers

It is generally believed that Wrynecks, like many other migrant birds that have summered in northern Europe, fly non-stop to southern Europe or North Africa where they then break their journey to refuel before crossing the Sahara, possibly also making stopovers in the desert and beyond it (Barlein 1988; Hedenström and Lindström 1990). Little is known about stopover sites in Africa, but there are presumably many. It is likely that weather conditions during migration dictate when, and for how long, such stopover sites are used (Zwarts et al. 2009).

Yet, in Russia, it has been observed that some birds move short distances locally before migrating proper, perhaps stopping to moult or feed-up

FIGURE 11.6 Some Wrynecks from mainland Europe winter in the Mediterranean, others move on to Africa. This one on Isola Ventotene, Italy, in November 2020, was probably only passing through (PH).

(Iovchenko and Kovalev 2005). If food is abundant migrants will also stay for several days (and possibly longer) at stopover sites across Europe before undertaking a subsequent longer flight (Langslow 1977). Clearly, without resting and feeding, many birds will be unable to complete their journeys between breeding and wintering areas. Birds which fly over the Mediterranean may drop into islands, big and small. Many Italian islands, for instance, are known to be regular stopover sites, particularly in autumn (Spina and Volponi 2008).

Situated at the southernmost tip of Israel on the Gulf of Aqaba (the northern tip of the Red Sea), Eilat is one such stopover. To the south and south-west lies over 3,000 km of mostly continuous desert across the Sinai Peninsula and the Sahara, to the north is the Syrian Desert and to the east the Arabian Desert. These are all daunting arid areas for migrating birds to negotiate. In a study at Eilat that ran from 1983 to 2010, an average of 18% of the Wrynecks caught each spring were re-trapped: a high figure in comparison to other species. The average stopover for these birds was calculated to be five days and the birds were found to have significantly gained weight and were in much better condition than when first caught (Yosef and Zduniak 2011).

Fat reserves

Birds that migrate across enormous expanses of desert or sea need large fat reserves to sustain them. Indeed, some distinctly well-fed Wrynecks have been trapped at ringing sites prior to their journeys (Bairlein 1988). A study of the body masses and fat deposits of Wrynecks caught at the Ottenby Bird Observatory in southern Sweden found that in autumn their relative level of fat was higher than for songbirds that were undertaking comparable journeys. The fat levels of birds trapped in spring, however, were on average lower than in autumn. This may have been because individuals trapped in spring had covered longer distances than autumn birds (Hedenström and Lindström 1990).

On the Isle of May, Scotland, five birds that were ringed and then re-trapped after three to six days, were found to have rapidly gained weight. The increases in weight varied from 8.0 g to 11.4 g (mean 9.1 g) per bird, which means an average rise of 29.3 ± 2.5% in weight. The most rapid rate of increase averaged 3.0 g per day for three days, while the greatest increment of 11.4 g (38.5%) in weight was achieved in just four days. It was presumed they were feeding on ants (Langslow 1977). An individual trapped on 12 October 2017 at the Copeland Island bird observatory in Northern Ireland weighed a hefty 51 g and had a fat score of 7 (on a scale of 1–8). Yet not all migrating Wrynecks do so well. It is not unusual for researchers to trap emaciated birds or find dead ones, possibly as a result of a local paucity of food (Forrester and Andrews 2007). Scebba and Lövei (1985) found that Wrynecks wintering on the Italian island of Vivara gained little if any weight, and indeed did not need to fatten-up as they were not about to embark on onward long-distance flights. Nevertheless, some birds there apparently did not spend the whole winter on the island.

Ultimately, it is difficult to draw any firm conclusions on what the ideal weight is for a migrating Wryneck, as examined birds can show great differences. For example, birds trapped in autumn (from 24 September to 1 November 1985) at Oued Fergoug in northern Algeria ranged between 28.8 g and 48.7 g (Bairlein 1988). Another study, of spring migrants in the Camargue, France, found an even wider range, from 22.5 g to 53.8 g. Such variations may be due to a mixture of lean new arrivals and birds which have already fattened-up in preparation for the next leg of their journey (Glutz and Bauer 1994).

Extralimital records

Being migratory, Wrynecks are prone to vagrancy. The following are a few examples. There is one record for the Cape Verde Islands on 24 September 2018. Madeira has had fewer than five records in the last 50 years. Wrynecks are more regular on the Canary Islands, with most found in April on Fuerteventura, the closest island to Africa, and a single modern record from La Palma on 7 April 1994 (Martín and Lorenzo 2001). Further north, they are more regular in Ireland with up to ten found each year, mainly in August to October, often at ringing stations on Cape Clear and Great Saltee island. They are rarer in the north, recent records being single birds on Arranmore and Tory, islands off County Donegal, both in October 2014, and another on 12 October 2017 at Copeland Island bird observatory off County Down. Iceland has fewer than 20 accepted records, mostly in September. The species was observed on the Faroe Islands just 25 times prior to 2009, mostly from late August into September with some in May and early June, but none have been officially recorded since. Records from North America include a dead specimen at Cape Prince of Wales, Alaska, on 8 September 1945 (Bailey 1945) and two living birds seen, one at Gambell, St Lawrence Island, Alaska, from 2 to 5 September 2003 (Lehman 2005) and one on San Clemente Island, California on 25 September 2017 (Singer et al. 2020). One found in a mummified state on a military base in Indiana in February 2000 was presumed to have arrived already dead in a shipping container (Dunning et al. 2002). In Asia, Wrynecks are extremely rare vagrants to islands such as Sri Lanka, where there are just two accepted records, both in the Yala National Park in the Dry Zone, the last in December 2009.

Chapter 12

Breeding

Wrynecks are essentially monogamous. Records of polygamous breeding, such as a male with several females (polygyny) or a female with several males (polyandry), are difficult to verify and probably rare. A German study that used DNA fingerprinting and molecular sexing methods to investigate paternity and sex allocation, found an exceptionally low rate (0.68%) of extra-pair paternity (Wink et al. 2011).

FIGURE 12.1 A clutch of twelve typical plain white Wryneck eggs. Item 03875 in the collection of the Mátra Museum of the Hungarian Natural History Museum (GG).

Unlike most other woodpeckers within their range, which are generally sedentary and may maintain a loose attachment through the winter before breeding anew in the spring, the Wryneck's pair bond probably only lasts for one season. This is perhaps unsurprising given that most are migratory, and mates do not maintain contact when on passage and in wintering quarters. On the other hand, fidelity to breeding areas, even exact nest sites, seems strong. Many adults are thought to return to the place where they bred the previous year (Cramp 1985). Hence, a pair may reunite in successive years, despite having been separated, although this might be due to site-fidelity rather than mate-fidelity. It is also thought that most birds which fledged the previous year return to their natal sites to breed in the following spring (Demongin 2016). Then again, there is also evidence that breeding populations have an open recruitment system in which a significant number of breeders are individuals that were not raised there but arrived from elsewhere (see Chapter 7). Some of the birds produced by these breeders also subsequently disperse to other populations (Schaub et al. 2012).

Home range

The sizes of home ranges vary greatly and are hard to define. They are likely to be related to the availability of resources, foraging sites and nesting locations (Mermod et al. 2009). In general, they are large at the start of the breeding season but shrink after pairs are formed. For example, a study of seven radio-tracked individuals in Switzerland found sizes ranging between 2.1 ha and 9.2 ha (Weisshaupt et al. 2011). A radio-tracked bird in Belgium roved over 255 ha before nesting but then just 8 ha when it was incubating eggs and feeding chicks (Kervyn and Xhardez 2006). In the Netherlands one pair occupied 45.5 ha of a forest clear-cut area and adjacent heath (based on places from which the birds called), but of that only 4.4 ha was used for foraging (Bijlsma 2014).

Ultimately, the quality and habitat diversity of an area dictates whether Wrynecks breed in it and influence home range size. In high-quality environments, the density of pairs will invariably be higher than in those of lower quality (Schaub et al. 2012). Numerous factors clearly influence the number of pairs in any given area, but where good nesting possibilities and abundant ant prey combine, densities can be high. For instance, studies in Italy have found 1–11 pairs per km^2 (Brichetti and Fracasso 2020). In prime areas nests can be close to one another, less than 100 m apart, exceptionally even 20–50 m, without serious conflicts ensuing (Glutz and Bauer 1994). Nonetheless, care should be taken in estimating densities without confirming occupied nests as both sexes vocalise and also often switch singing locations which can create an impression that more birds are present than is actually the case (pers. obs.).

Finding a mate

Soon after they arrive in breeding areas, Wrynecks begin to call to attract a mate (see Chapter 6). They call from exposed perches, often on the tops of bushes, from tree snags, fenceposts and sometimes from wires. It seems that both sexes search for nest cavities and show potential nest sites to prospective mates. Later, when the pair is formed and the nesting cavity selected, just one regular singing post close to the site tends to be used.

Courtship

Wrynecks display close to their nest hole. They perch close to each other, sway their heads rhythmically, stretch the neck, point their bill upwards, spread the tail and wings, and chase each other in flight (Menzel 1968; Löhrl 1978). Once a bird has chosen a cavity, it will repeatedly enter it, sometimes uttering a muted song or making soft calls and occasionally tapping excitedly at the entrance or chamber walls (see Chapter 6). This is probably done to confirm ownership but also to entice its mate to inspect it. Males will also feed females during courtship and this may continue into the incubation period. Once paired, birds typically greet each when they meet, for example, at nest change-overs, by raising their crown feathers, ruffling their body plumage, swaying their heads and stretching their necks, often in unison, and uttering soft notes in duets (Menzel 1968).

Copulation

Flitting flights may precede mating which takes place either on the ground or on a bare branch (Cramp 1985), usually close to the chosen nest site. Unlike most other woodpeckers, this may be performed crosswise on a branch rather than lengthwise. The female crouches, soliciting the male to mount her, and the male may fly directly onto her back or hop on if already beside her. The male cocks his tail up or places it to the side and shrouds the female with his drooped wings, which quiver during the act. He may also dart his tongue in and out and peck his mate's nape. Several copulations per day take place (Ruge 1971.) Reverse mounting, where a female mounts a male to solicit copulation,

FIGURE 12.2 A recently arrived bird belts out its song to declare its presence and attract a mate. April 2021, Novo Yankovo, Bulgaria (DG).

may occur, as it occasionally does in other woodpeckers (Gorman 2020), but owing to the challenge of sexing the species, it is difficult to confirm.

Eggs

Clutches are laid anytime from April to mid-June. It may be 30 days after pairs arrive in the breeding area before egg-laying takes place, as courtship and pair formation must first be completed, and a suitable nest site found and occupied. Single eggs are laid at daily intervals (Cramp 1985). As is typical for many cavity-nesting birds, which lay their clutch at the bottom of a dark chamber, eggs are plain white, smooth but not necessarily glossy. Any markings are the result of being soiled.

FIGURE 12.3 A clutch of ten eggs laid in a nest box on the remains of a Great Tit nest. June 2021, Börzsöny Hills, Hungary (GG).

Size and shape: Woodpeckers in general have smaller eggs than most other birds of comparable size (Winkler et al. 1995). This is valid for Wrynecks, one study finding they produce eggs with a relative weight as a percentage of female body mass of 6.5% (Hogstad 2006). Yet, in the same study, total clutch weight as a percentage of body mass, calculated to be on average 59%, was higher than that of the true woodpeckers which have an average of around 34%. Eggs are sub-elliptical to oval in shape (Harrison 1974). In 34 clutches measured in Croatia, length ranged from 17.7 mm to 21.9 mm (average 20.1 mm) and width 14.2 mm to 15.9 mm (average 15.1 mm). The elongation index ranged from 1.19 to 1.43, an average of 1.33 (Dolenec 2020). Measurements taken of collections in the Czech Republic (Mlíkovský 2006) and Hungary (Solti 2010) produced similar results.

FIGURE 12.4 Recently hatched chicks: blind and bald. June 2021, Lipník nad Bečvou, Czech Republic (ZA).

Clutch size: Wrynecks generally lay more eggs than other woodpeckers (Hogstad 2006). The size and volume of their eggs does not, however, seem to influence clutch size (Dolenec 2020). Between 6 and 12 are typically laid. Numerous studies across Europe, both in the field and of clutches in collections, have produced averages of 8–10 eggs (Linkola 1978; Solti 2010; Hrabovský 2013; Brünner and Rödl 2018; Dolenec 2020). In a German nest box study, where two clutches per year were the norm, first clutches had an average of 9.9 eggs and second ones 7.7 (Becker and Tolkmitt 2007). Sometimes exceptionally large clutches of 20 or more are found, but this is likely the result of nest parasitism (egg dumping) by extra-pair females (Fraticelli and Wirz 1991). Such clutches are probably not successful as broods of such sizes are not found. Wrynecks are indeterminate egg layers, that is, if their eggs are removed, the female will replace them consecutively. Experiments suggest they may be prolific layers. Alderson (1897) conducted several trials, removing one egg each day as soon as it was laid, and in one recorded a remarkable 62 eggs. In a similar experiment Warga (1926) recorded 33 eggs laid in 33 days.

Incubation

Compared with other secondary cavity nesters, Wrynecks have a relatively short incubation period, typically 12–14 days (Hogstad 2006). Both sexes incubate and usually begin once the clutch is complete, occasionally before the last egg is laid. Parents exchange duties 3–5 times daily (Ruge 1971). The sitting bird is silent and its mate outside the nest is also generally quiet and inconspicuous at this time, so it can appear, erroneously, that a previously vocal and obvious pair are not breeding and have abandoned the area (pers. obs.). Due to the difficulty of sexing birds, determining which parent carries out overnight incubation is impractical. In most woodpeckers, however, it is the male (Gorman 2014).

Hatching and brooding

Whatever the clutch size, it is seldom that all eggs hatch, but hatching success rates of up to 75% are in line with those of other woodpeckers (Becker and Tolkmitt 2010). Hatchlings are altricial, naked, blind, deaf, and typically weigh less than 2 g. One or two often perish soon after hatching (Mulhauser and Zimmermann 2014). Of those that survive, one or two may be runts (Hrabovský 2013). A study in Finland found that pairs raised on average 5.5 young per nesting after such losses (Linkola 1978). Nestlings open their eyes

FIGURE 12.5 A typical huddle of nestlings, which are 9 to 10 days old. Note the differences in size between the siblings. June 2006, Milovice, Czech Republic (JC).

FIGURE 12.6 Nestlings at 14 to 15 days old. Note that they tend to face towards the walls of the nest box. June 2021, Balatonmáriafürdő, Hungary (GG).

at around 8–9 days (Sutter 1941). They form a compact huddle, climbing onto each other to form a pyramid, as many woodpeckers do. As they grow, they become more restless and make looser formations. They develop rough 'heel' pads on the backs of their legs which help them clamber up the chamber walls (Harrison 1974). Nestlings are brooded constantly by both parents until about five days old, after which intensive feeding takes preference (Menzel 1968).

Feeding the brood

Both parents feed their nestlings but as already mentioned they are not sexually dimorphic, so it is difficult to judge the ratio of visits by each. In exceptional circumstances one parent may raise the brood. A radio-tagged adult, believed to be a male, in the Ardennes, Belgium, was found to be incubating a clutch and then feeding the brood alone. However, only two chicks hatched from five eggs (Kervyn and Xhardez 2006). An intriguing observation in Switzerland

FIGURE 12.7 A parent arrives with food for its hungry chicks, one already waiting at the nest entrance. June 2020, Modena, Italy (ET).

concerned Wryneck nestlings in a tree cavity being fed by a female Lesser Spotted Woodpecker (Contejean 1998). Another case, also in Switzerland, involved a pair of Hoopoes that had lost their own brood feeding Wryneck nestlings with Mole Crickets *Gryllotalpa gryllotalpa* while their actual parents continued to feed them ants. Interestingly, the growth rate of the nestlings was found to be lower than average, although three out of the eight nestlings still fledged (Mermod et al. 2008). For more on feeding young see Chapter 14.

Nest sanitation

Nestling droppings, enclosed in a mucous faecal sac, are removed by parents (Klaver 1964). They are not usually taken away when the chicks are small, possibly being consumed by the parents, but from about seven days are carried away in the bill, not dropped below the nest. Later, they may once again be overlooked, probably as the nesting chamber becomes too crowded for the adults to enter (Ruge 1971). As feeding intensifies and the nestlings grow, droppings are taken away after most feeding visits (pers. obs.). Dead chicks are also usually removed (Menzel 1968). Eggshells are seldom taken away, being consumed by the chicks or even fed to them (Löhrl 1978). For more on this see Chapter 14.

FIGURE 12.8 Parents try to keep the nest clean by taking away chicks' droppings which are enclosed in convenient 'wryneck nappies'. June 2018, Traisen, Austria (TH).

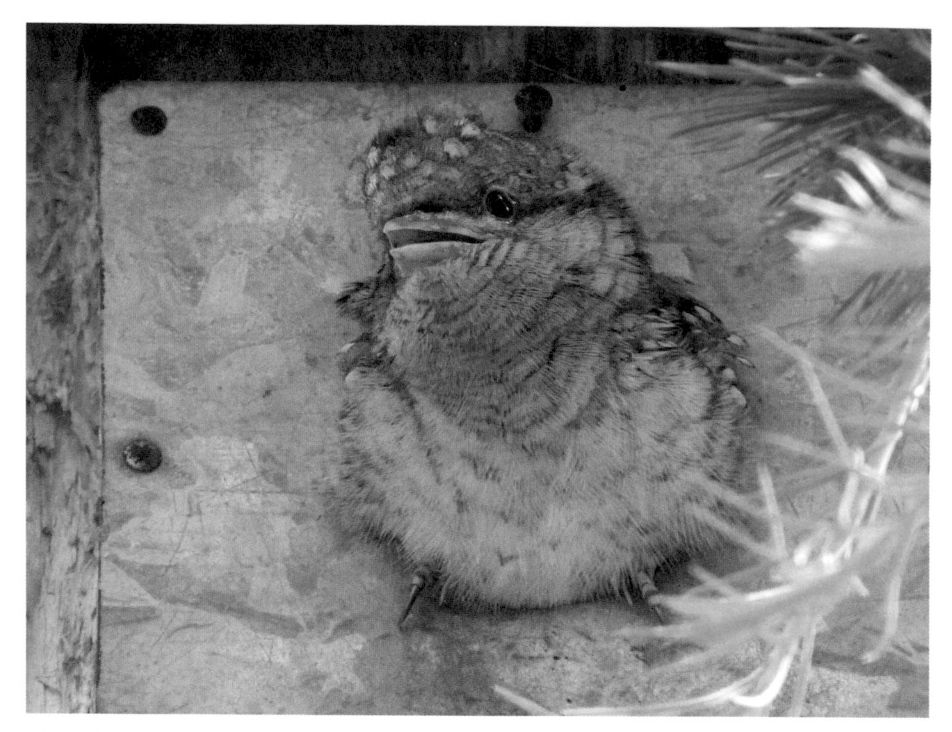

FIGURE 12.9 This well-fed chick will soon fledge and leave the relative comfort and safety of the nesting cavity. June 2020, Jihlava, Czech Republic (AT).

Fledging

Nestlings start to look out from the nest entrance when about 18 days old (Bijlsma 2014). When peering out they occasionally dart their tongue in and out of their bill (Bussmann 1941). They typically fledge at 18–24 days after hatching (Ruge 1971; Freitag et al. 2001; Hogstad 2006). Hence, the total period from the start of incubation to fledging is 30–8 days. It may take several days for all of them to leave the nest. They are prompted to do by their parents which reduce feeding visits and by the increasingly cramped conditions in the chamber, which results in siblings pushing and pecking each other. Fledglings are fed outside the nest for up to three weeks, but after that, unlike most other woodpecker young, they tend to leave the immediate vicinity of the nest and become independent of their parents. Such comparatively rapid independence may be because Wrynecks regularly raise two broods.

Breeding success

Breeding success in Wrynecks is generally high. It is influenced by various factors: fitness and experience of the pair, food availability, habitat quality, location of nesting cavity, weather and predation can all play roles. They often lay large clutches, sometimes double figures, but the corresponding number of chicks rarely fledge (Hrabovský 2013). Brood sizes are usually sufficient to maintain a stable population despite any losses in the egg to fledging period. A long-term Swedish study compared the sizes of 1465 broods over the course of four decades (two periods, 1962–81 and 1982–2001) and involved the ringing of 10,031 young (Ryttman 2003). It was observed that although the number of broods found in the second period (449) was far fewer than in the first (1016), breeding success within broods did not decrease; in fact an increase from 6.7 to 7.1 young per brood was recorded. It was concluded that the species, despite apparently high losses when the number of eggs laid is considered, can maintain their population size with a production of about 5.5 young per nest. In other studies, an average productivity of 3 to 4 young per pair and a fledging rate often over 70% has been found (Becker and Tolkmitt

FIGURE 12.10 A chick scrambles out of the nesting cavity for the first time. June 2021, Lipník nad Bečvou, Czech Republic (ZA).

FIGURE 12.11 A newly fledged chick and a sibling that will soon join it outside the nest. June 2021, Lipník nad Bečvou, Czech Republic (ZA).

2010; Brünner and Rödl 2018). In the latter study, of 41 broods in nest boxes in Bavaria, a particularly impressive success rate of 78% (based on at least one chick surviving) with an average of 7.4 young fledging was documented. Indeed, unless a predator accesses the nest, which usually results in the whole brood being lost, nestling mortality is generally low. Another German study recorded just 142 dead chicks from a total of 3024, with a mean of just 4.7% (Becker et al. 2014).

Sudden changes in weather can impact nestling survival. Two weather events in particular, falling temperatures and heavy rainfall, are often significant. Yet it seems that drops in temperature are less damaging than downpours. A Swiss study (Geiser et al. 2008) determined that it was how ants reacted to weather events, rather than the condition of nestlings, that mattered. In periods of heavy rain ants go deeper underground and are therefore more difficult to find. Wrynecks respond to this by adjusting their foraging and feeding strategies. They reduce visits to their nestlings, which probably means slower nestling growth in the short term, but when conditions improve, compensate by significantly increasing their activity and their young subsequently recover.

FIGURE 12.12 A fledgling on its first day out of the nest. June 2012, Noszvaj, Hungary (GG).

Yet it is likely that this can only work if the fall in available food is temporary, and the area hosts high numbers of ants. The timing of such weather events is also critical as older nestlings will most likely be able to cope with a reduction in food better than recently hatched ones. Indeed, the authors concluded that nestling survival rates increased with age. Still, it is worth noting that in degraded habitats with poor numbers of ants, this ability to endure variable weather will likely diminish, particularly in fringe Wryneck populations.

Of course, some nesting attempts do not succeed at all. Failures occur for several reasons, such as predation (see Chapter 15) and even competition between neighbouring pairs when they throw out each other's clutches (Becker and Tolkmitt 2010; Becker et al. 2014). Although an extremely rare occurrence, there are records of Wrynecks failing to breed owing to brood parasitism by Common Cuckoos *Cuculus canorus*. Incredibly, as they were inside cavities, Cuckoo chicks and eggs have been documented in Wryneck nests (Makatsch 1955). An astonishing record from Romania in May 1926 mentions a nest hole in a tree 1.6 m above the ground that contained nine Wryneck eggs, one Cuckoo egg and one Marsh Tit *Parus palustris* egg (Dobay 1927). With respect to those observers, the possibility that the supposed Cuckoo eggs were in fact aberrant Wryneck eggs cannot be ruled out. Reports of Cuckoo chicks are more difficult to dismiss.

Extra broods

If a clutch is lost replacement clutches are usually laid. A study in Germany's Harz Mountains found that of 664 broods 76 were replacements (Becker et al. 2014). Furthermore, even when a first breeding is successful, Wrynecks often raise a second brood in the same season (Becker and Tolkmitt 2007; Schaub et al. 2012). Occasionally, three broods are attempted, especially in southern Europe where the days and season are longer (Glutz and Bauer 1994). At northern latitudes, for example in Scandinavia, second broods are rare (Ryttman 2003). Climate and prey abundance undoubtedly play a part in extra brood attempts. During favourable conditions, a first clutch can be laid early, and chicks fledge swiftly, with pairs then able to make a second breeding effort (González et al. 2002). Such broods are usually attempted by the same pair, mates seldom changing (Schaub et al. 2012). Nonetheless, a study of 50 broods in Germany found three cases of males starting a second with a different female, while nestlings of the first had still not fledged (Wink et al. 2011). These males raised 13 young per season, compared to 6.14 for monogamous males with single broods. In another German study considerable differences in success between successive broods were noted, with the first being the most successful (Becker and Tolkmitt 2010). In a later study Tolkmitt et al. (2020) found that first broods were more likely to be recruited into the population than second or replacement broods which raises a question of fitness in the latter two. Replacement and second broods may be raised in the same cavity as the first or in another nearby (Bijlsma 2014).

FIGURE 12.13 No time to lose when raising a brood. The rapid change-over of parents eager to feed their offspring. June 2021, Ipolydamásd, Hungary (GG).

Chapter 13

Cavities

All woodpeckers nest in cavities, the majority of species excavating them in trees for this purpose. These species are called 'primary cavity-excavators', while those that do not excavate their own cavities are termed 'secondary cavity-users'. Wrynecks are one of the few woodpeckers that are secondary cavity-users, that is they rely on natural holes and hollows in trees and stumps, and old woodpecker cavities. To a lesser extent they will also nest in burrows made by other birds in earthen banks, and holes in walls and buildings. Wrynecks also readily occupy nest boxes. Though they never excavate a complete cavity, they may make minor modifications to existing ones, slightly widening or deepening them, by light pecking. Though rare, cases when nesting sites previously used by other species were significantly reworked have been reported (Åbro 1962). As their bill is relatively weak, certainly when compared with other woodpeckers, they are only able to do this in soft or rotten wood (Varga 1981).

Nesting cavities

The importance of cavities for nesting cannot be overestimated. Wrynecks do not require extensively wooded areas to breed in, what is crucial is an environment that contains a good number of trees with suitable cavities. The availability of prime sites for nesting, and perhaps roosting, is a factor that influences their distribution. Indeed, the importance of an adequate supply of cavities has been shown experimentally using nest boxes (Zingg et al. 2010). In any given area some available cavities will be unsuitable owing to being too shallow, damp or infested. When a Wryneck finds a potential site, it will show it to its mate by going in and out and tapping around the entrance (see Chapter 6). Tree hollows and holes made by Great Spotted, Syrian *Dendrocopos*

FIGURE 13.1 A bird assesses a hollow in the cut limb of an old walnut tree as a possible nest site. April 2021, Börzsöny Hills, Hungary (GG).

syriacus, Middle Spotted *Leiopicus medius*, Lesser Spotted (Pakkala et al. 2019) and Eurasian Green Woodpeckers *Picus viridis* (Bijlsma 2014) are all used in Europe and elsewhere across the range those of other woodpeckers.

Burrows in sand or earth banks made by Common Kingfishers, Sand Martins *Riparia riparia* (Alter 2002) and European Bee-eaters *Merops apiaster* (Schulze 2008) are also occupied. On the southern Siberian steppe where there are few if any trees, Wrynecks nest in the burrows of ground-living mammals such as sousliks and pikas. A curious observation in Switzerland involved a nest in a hollow in an old chestnut tree that had two entrances. Both parents usually entered through the upper entrance and left via the lower one some 20 cm below (Sermet 2001). An even more astonishing record from Hungary documented a pair breeding in an old mitten that had been hung up to store garlic (Chernel 1903).

FIGURE 13.2 An old Great Spotted Woodpecker hole in an apple tree used as a nest site. June 2021, Ipolydamásd, Hungary (GG).

FIGURE 13.3 An old Middle Spotted Woodpecker hole in a cherry tree used as a nest site. June 2012, Noszvaj, Hungary (GG).

Cavities in the walls of cabins and sheds, under house eaves and even in log piles have also been recorded (Cramp 1985; Schaub et al. 2012). A report from the Czech Republic involved a pair which successfully nested in a hole in a stone wall, despite there being many seemingly suitable tree cavities in the vicinity (Reiter 2013). It was thought the pair may have been motivated to use the wall site as House Sparrows *Passer domesticus* and Great Tits had previously nested there. On the other hand, there are also cases where Wrynecks use a cavity first, only for it to be used in subsequent years by other species (Nenadović 2008). For cavity competition see Chapter 15.

A study in vineyard-dominated landscapes in north-east Italy concluded that Wrynecks occupied those vineyards which had many potential nesting sites in the form of the hollow pipes used to support vines (Assandri et al. 2018). Nevertheless, every breeding attempt observed in these pipes failed during the egg-laying stage. The authors believed that although the pipes were obviously attractive to pairs, they proved to be 'ecological traps' as their internal temperatures were too high for the eggs to develop, often around

FIGURE 13.4 Nest in a vertical hollow in the trunk of a cut cherry tree. Despite being open at the top and exposed to the elements, the brood fledged successfully. Note the rather daredevil ant on the bird's bill. June 2018, Traisen, Austria (TH).

10°C higher than outside. It was surmised that inexperienced, first-time breeders were attracted to the pipes rather than more experienced, and perhaps dominant, birds which used the less numerous but more favourable sites such as nest boxes.

Orientation

As Wrynecks breed in existing cavities that they have not themselves excavated, the direction that the entrance hole of their nest faces may not be entirely of their choosing. As ideal cavities are often in short supply and in demand by other wildlife, it is likely that the species cannot be too particular in this respect. Whether they have preferred orientations and select nest sites from what are available locally with entrances facing in those directions is unknown.

Nest boxes

As mentioned in Chapter 9, a lack of suitable nesting sites can be a limiting factor in the breeding season. This problem can be partly addressed by placing out nest boxes, as Wrynecks will readily use those specifically designed for them, and those erected for other species. Indeed, they will evict sitting songbirds and accept boxes that were used earlier in the spring by others, after removing any nest material or debris. On the other hand, a study conducted in Croatia from 1985 to 2018, in which boxes were placed out for songbirds, found that only around one in every hundred was occupied by Wrynecks (Dolenec 2020). Whether this low number is due to there being few Wrynecks in those areas, the unsuitability of the boxes, their location, an abundance of other cavities, or other factors, was unclear. Conversely, nest boxes specifically erected for Wrynecks are also often occupied by other secondary cavity-users. For example, in Sweden researchers found that up to 80% of Wryneck boxes were used by other birds (Ryttman 2003). Boxes intended for larger birds are also used. In Switzerland, for instance, boxes with an internal area and entrance hole size to suit Hoopoes were regularly occupied by Wrynecks although they are much smaller birds (Schaub et al. 2012).

There is some evidence to show that nest boxes may even be 'better' than natural sites, as the reproduction rates of pairs using them can be higher. For instance, clutch and brood losses owing to flooding after heavy rain and predation are often lower in boxes than in other types of cavity (Zingg et al. 2010). In a study in Sweden, an increasing number of ringed broods was attributed to the increased provision of boxes (Axelsson et al. 1997). Indeed,

FIGURE 13.5 This 'luxury' nest box has a horizontal perch for the parents to rest on and a metal plate to prevent predators from widening the entrance hole. July 2020, Jihlava, Czech Republic (AT).

FIGURE 13.6 This box, which the Wrynecks used after evicting Great Tits, has a sloping roof and convenient perch. May 2021, Kocsér, Hungary (RP).

FIGURE 13.7 Box made from a section of log, hollowed out and fitted with a roof and floor. May 2021, Kocsér, Hungary (RP).

the probability of an area being occupied will often increase when boxes are placed in it. In areas where there are good foraging habitats but few tree cavities boxes can increase breeding densities to as high as one pair per hectare (Becker and Tolkmitt 2007).

Box designs and dimensions

Although Wrynecks use nest boxes intended for other birds, those constructed specifically for them tend to be more successful (Zingg et al. 2010; Brünner and Rödl 2018). Correct design and dimensions are important, although boxes do not need to be impeccable. Indeed, natural sites such as tree hollows and woodpecker cavities are often far from perfect and, as mentioned above, unusual sites are used when the options are limited. Various types are available commercially, but home-made ones work well. They can be made from wood, wood-concrete or a section of log, and be box-shaped or cylindrical. Colour is not important and they do not need to be camouflaged, but it is perhaps best if they are plain and drab in order not to attract unwanted attention. An ideal size is 20 × 20 cm (20 cm diameter for cylinders) with a height of 35–40 cm. They should be hole-fronted, not open-fronted, with an entrance

FIGURE 13.8 Simple but solid and effective 20 × 20 cm boxes. With a hinged roof to allow inspection and waterproofed with paint on the outside (AK).

hole diameter of 34–35 mm positioned about 5 cm from the top of the box. An entrance of this size prevents larger competitors such as Common Starling *Sturnus vulgaris* and Hoopoe using the box (Zingg et al. 2010) and limits access by predators such as Pine Martens *Martes martes*, Beech Martens *Martes foina* and Stoats *Mustela erminea*. Obviously, smaller birds can enter but in doing so they run the risk of being expelled and their chicks even killed by Wrynecks. Nest material does not need to be placed inside, but the bottom of the chamber should be slightly concave, or a little sand or sawdust added to cushion and prevent eggs from rolling around. Wrynecks do not line nests with material, although Menzel (1968) mentioned that they sometimes place pine needles into bare nest boxes. The roof can be rainproofed by being slightly sloped to facilitate run-off or boxes tilted slightly forward when they are erected. A hinged lid that opens enables access for ringing of chicks and cleaning after the breeding season.

FIGURE 13.9 Box location is, of course, important. This one is in a clearing, facing away from the forest towards an open pasture. Vértes Hills, Hungary (GG).

FIGURE 13.10 This box is in an open, grassy landscape where there is good foraging but few trees with suitable cavities for nesting. Börzsöny Hills, Hungary (GG).

Box location: Choosing the right location is vital. Boxes placed in good Wryneck habitat with adjacent dry, low, sparse vegetation will be more readily used and can even boost local populations. Pairs do not seem to be overly sensitive to disturbance, often using boxes placed in gardens and by roads and railway lines. Nevertheless, a relatively secluded spot, though not hidden in foliage, is prudent. A clear flight path to the entrance, for example, on the side of a tree facing away from woodland rather than towards it, is essential. Several boxes can be placed in one area, some tens of metres apart, to provide alternatives. It has been observed that areas with more than one box are more likely to attract the species (Zingg et al. 2010). Inevitably, some other wildlife (birds, small mammals, insects) will occupy Wryneck boxes, and this should be expected and tolerated. As already mentioned, whether Wrynecks prefer nest sites facing in certain directions is unknown, but as they are presumed to be essentially a warmth-loving species a southerly orientation and one aligned away from the prevailing wind direction is probably judicious. It does not seem to be problematic if the next box faces the midday sun (pers. obs.).

Height above ground: As Wrynecks do not excavate their own cavities, they have presumably had to evolve to accept what is available in any given area. This certainly seems to be true when it comes to nest site height, with pairs often using low locations, sometimes in tree stumps less than 1 m above ground level, and even in the ground itself. Consequently, boxes do not need to be placed high up; an entrance hole at between 1.5 and 3 m from the ground is ideal. High locations do not deter predators such as martens and snakes, as they can climb to any height.

FIGURE 13.11 The author placing a nest box at about 1.5 m above the ground. Wrynecks do not seem to be too concerned about the height of their nest sites. Pest County, Hungary (GG).

FIGURE 13.12 A pair of Wrynecks bred for several years in this log nest box, originally intended for tits, just 1 metre above the ground in a village garden. Szár, Hungary (GG).

Chapter 14

Foraging and Food

Woodpeckers globally can be placed in three basic groups according to their diet: omnivores, eaters of arboreal invertebrates, and species that feed predominantly on ants (Mikusiński and Angelstam 1997). The omnivores are the generalists, the least specialised in terms of foraging and feeding, and the other two groups are specialists. The ant-eaters are obviously those that are specialists in finding and preying upon ants, but there are innumerable kinds of ant, and these can, for convenience, be placed into two groups: arboreal and terrestrial. In terms of diet, Wrynecks are very much specialist birds, feeding mainly upon smaller species of terrestrial ant.

Diet

Wherever they occur, but especially in the breeding season, Wrynecks need ants to be abundant and easy to get at. This has been well studied and summarised, for example in Cramp (1985), Menzel (1968), Ruge et al. (1988), Glutz and Bauer (1994) and by others since. Indeed, across the breeding range, the diet of the species is often composed entirely of ground-living ants, which are taken at egg, larval, pupal and imago stages. Other invertebrates are eaten but are probably not especially sought and are consumed only opportunistically, when locally abundant, or when ants become scarce, as during bad weather. They include aphids, flies, small beetles, woodlice, caterpillars, grasshoppers, cicadas, craneflies, mayflies, moths, butterflies and spiders, which are also eaten in all stages of their development (Cramp 1985; Bijlsma 2014). When moving along coastal areas and islands on migration, Wrynecks often forage among rocks and seaweed on beaches, presumably searching for fly larvae, sandhoppers and the like. In African wintering quarters, termites also are probably taken. Nevertheless, most studies that have examined the

FIGURE 14.1 An adult with a bill full of ant pupae for its chicks. June 2018, Traisen, Austria (TH).

FIGURE 14.2 Stomach of a Wryneck and its contents, the remains of Black Garden Ants *Lasius niger*. Sample from the Linz Biologiezentrum, Austria (SW).

contents of Wryneck droppings (and regurgitated pellets) have found them to contain mostly chitinous parts of ants, the remains of other insects usually being negligible (King and Speight 1974; Terhivuo 1976; Benabbas-Sahki et al. 2015).

Ants

Ant species taken by Wrynecks in Europe include *Tapinoma erraticum*, *Formica rufa*, *F. polyctena*, *F. fusca*, *F. cunicularia*, *F. rufibarbis*, *F. exsecta*, *Lasius niger*, *L. alienus*, *L. emarginatus*, *L. flavus*, *L. platythorax*, *Leptothorax unifasciatus*, *Myrmica lobicornis*, *M. sabuleti*, *M. scabrinodis*, *M. schencki*, *M. sulcinodis*, *M. rubra*, *M. ruginodis*, *Tetramorium caespitum* and to a lesser extent *Crematogaster* and *Camponotus* species. Most studies have found ants in the *Lasius* genus to be the commonest items in the diet (Bitz and Rohe 1993; Freitag 2000; Mermod et al. 2009; Weisshaupt et al. 2011; Bijlsma 2014). An analysis of droppings collected in Britain discovered that they were composed almost entirely of fragments of workers and cocoons, the content very much

FIGURE 14.3 Swarming Meadow Ant colony (possibly *Formica pratensis*). Such ground-living ants are a favourite food. April 2021, Vértes Hills, Hungary (GG).

like those of Eurasian Green Woodpecker (King and Speight 1974). In Algeria, faeces collected in winter contained mainly *Tapinoma nigerrimum*, with smaller numbers of *Pheidole pallidula*, *Tetramorium biskrensis* and *Plagiolepis barbara*, interestingly mostly in adult form (Benabbas-Sahki et al. 2015). Analysis of droppings and pellets in Hokkaido, Japan, identified 13 ant species in the four genera *Formica*, *Myrmica*, *Pheidole* and *Lasius*, with *Lasius japonicus* being the most frequent (Yoshimura et al. 2003). The particular ant species consumed are, of course, influenced by location, availability and season.

Foraging niche

This species is not unique among woodpeckers in foraging mainly on the ground. Several species throughout the family's range, including sympatric ones such as those in the *Picus* genus in Eurasia and some in the *Campethera* genus in Africa, also do so (Gorman 2014). The Wryneck's preferred foraging habitat differs regionally and locally but, overall, open terrain with barren ground interspersed with vegetation is typical – in short, places where ant colonies are concentrated. For Wrynecks breeding in orchards and vineyards

FIGURE 14.4 Bare ground such as this, which has been heavily browsed by rabbits, is often good foraging habitat. April 2010, Norfolk, England (NeB).

or on other cultivated land, open ground for foraging is considered to be vital (Weisshaupt et al. 2011). Studies have produced figures ranging from 30% to 50% (Mermod et al. 2009; Coudrain et al. 2010) and to up to 80% (Schaub et al. 2010) of bare ground as being ideal in any given area. Research in the Netherlands found that Wrynecks did not look for food in places where ground vegetation cover exceeded 85%, as colonies of their preferred ants were usually scarce in such habitats (Bijlsma 2014). This is no doubt why pairs that nest in wooded areas, where ground cover can be high, are often seen foraging at the edges of dirt tracks and roads. Similarly, pastures are preferable to meadows, because the latter, although often hosting many ants, seldom include patches of exposed ground (Weisshaupt et al. 2011). Areas totally devoid of vegetation, however, are not ideal, either, as ants tend to become less numerous in such places. Nonetheless, Wrynecks will forage in lush and high grass, and in crops such as lucerne when ants are plentiful there (pers. obs.). Ultimately, the best foraging habitats seem to be mosaics typified by patches of both barren and verdant ground, places where ants are both abundant and, importantly, easy to access (Coudrain et al. 2010).

Foraging methods

For the most part Wrynecks collect prey directly from the ground with their tongue. As the tongue lacks barbs, prey is glued to its sticky surface rather than impaled. When extended out of the bill, the tongue is used to probe and catch ants from within their colonies just below the surface or from under stones. Wrynecks will also dig into ant nests with their bill but, as they can penetrate only a little way, many ants deeper in the colony will remain out of reach (Bijlsma 2016). Prey is probably detected visually by the birds as they hop on the ground but is sometimes spotted from low perches in bushes and trees (pers. obs.). Ants on or just under the bark of trees are also gleaned, and Wrynecks will peck and probe though do not hack deeply into wood for prey. Walls, paving stones and rocks are also investigated. Productive feeding spots are returned to repeatedly. Occasionally, inventive ways of foraging are employed. For instance, a bird will wait motionless in one spot, such as on a stone, and dart out its tongue to ensnare ants as they file by (pers. obs.). Individuals have also been observed using their tongue to catch flies and to take mayflies from water surfaces by perching on overhanging branches and 'angling' with it (Cramp 1985).

FIGURE 14.5 A Wryneck uses its long sticky tongue to capture prey directly from the ground. April 2010, Norfolk, England (NeB).

Feeding young

When feeding their chicks, Wrynecks often forage in the immediate vicinity of the nest, sometimes as little as 40–50 m. Distances covered, however, depend on prey availability and abundance; for example, in a clear-cut forest in Belgium a radio-tracked individual roved between 150 m and 500 m from the nest in search of food (Kervyn and Xhardez 2006).

Parents tend to bring food separately but sometimes they arrive simultaneously, one usually waiting outside the cavity until the first leaves. When nestlings are tiny both parents may enter together or, occasionally, one will pass food to its mate at the cavity entrance who then feeds their chicks. Sometimes, one bird is dominant, driving its mate off and feeding the chicks first, the second parent only returning when it has finished (pers. obs.).

After around four days feeding intensifies (Geiser et al. 2008). In a high-tech field study in the Rhône plain, Switzerland, Passive Integrated Transponder (PIT) technology was used to monitor feeding activity at two nest boxes (Freitag et al. 2001). Adults were tagged with a PIT and a reader placed by

FIGURE 14.6 Parents usually arrive independently with food for their nestlings. June 2011, Szár, Hungary (AK).

FIGURE 14.7 Occasionally a pair will arrive with food for their nestlings at the same time. May 2021, Kocsér, Hungary (RP).

FIGURE 14.8 A parent pauses cautiously on the top of its nest box before dropping down to enter to feed its nestlings. June 2021, Balatonmáriafürdő, Hungary (GG).

the box entrance. On average 150 to 190 feeding visits per day, sometimes over 200, were made. On some days over 14 hours of feeding took place, with activity highest in the morning hours. It was found that feeding efforts varied on a daily basis, as prey abundance and weather, particularly temperature and precipitation, all exerted an influence. The number of visits fell when it was too wet, cold or hot, as all of these conditions affected prey.

Ants, when abundant, may comprise almost all of the food given to chicks (Glutz and Bauer 1994). A study of seven nests in Switzerland, which included more than 2,000 observations, found that ants were brought to nestlings in 95% of cases (Freitag 2000); the majority were *Lasius*, *Formica*, *Tetramorium*, *Tapinoma* and *Myrmica* species, mostly larvae and pupae, which indicated that they had been taken directly from the insects' nests. Indeed, in the nesting period ants are mainly collected directly from their colonies, especially larvae and nymphs which dwell in the upper levels of colonies in order to take advantage of the sun's warming rays. When it rains, or the temperature significantly drops, ants move their offspring deeper underground (Freitag 1996). Similarly, if the temperature in the upper levels of a colony is too high, as in midday and afternoon hours, these insects will also shift their broods deeper to avoid overheating. This might also explain why Wrynecks tend to make fewer feeding visits during afternoons. Another factor that shapes feeding activity is brood size. Bussmann (1941) asserted that the larger the brood, the higher the rate of feeding, and that feeding intensifies in the second half of the growing period. The individual foraging skills and experience of parents probably play a role, too.

FIGURE 14.9 When the chicks are large parents typically cock their head sideways to force food into their open bills. Perhaps to avoid being stabbed! June 2020, Modena, Italy (ET).

Parents may fly directly to the nest entrance and enter at once, or they may alight and pause briefly there before entering. They may also land nearby and quietly survey the area, sometimes for several minutes, before hopping to the entrance. Similarly, they may exit the nest and fly away immediately, or pause at the nest entrance and lick their bill clean of any food that has stuck to it, or alight nearby before flying on (pers. obs.). Some individuals, even within a pair, seem to be more cautious than others. Local circumstances no doubt also play a part.

At first, parents enter the nesting chamber completely. They do this for up to 18 days but later, when the chicks are large, they hang down into the crowded chamber from the entrance (Ruge 1971). When small, nestlings have flesh-coloured gape flanges which parents touch to stimulate them to open their bills (Klaver 1964). Food is brought in the bill and is usually so copious that it is visible. It is inserted directly into the chicks' throats in thrusting and pumping movements (Sutter 1941).

Unusual food

As already outlined, Wrynecks have a highly specialised diet, but they will occasionally take advantage of other sources of food. Snails (Terhivuo 1976), Yellow Slugs *Limacus flavus* (Allen 2004) and even the tadpoles of Common Frog *Rana temporaria* (Cramp 1985) have all been recorded as being fed to chicks, but these types of food are exceptional and do not play a regular or important role in their nutrition. Some woodpecker species (mostly in the New World genus *Melanerpes*) are known to prey frequently on the eggs and nestlings of other birds. Within the range of the Wryneck, Great Spotted Woodpeckers often do this, and Syrian Woodpeckers occasionally do so (Gorman 2004). Although they regularly remove eggs and chicks from already occupied nest holes (see Chapter 13), few Wrynecks seem to prey specifically on the clutches and broods of other birds. They may on occasion eat such food, but this, rather than being deliberate predation, is in most cases probably a consequence of the birds visiting cavities to inspect them for possible use and finding this convenient resource, as they do not raid the nests of non-cavity-breeding birds.

An astonishing observation from the Sundarbans, in north-east India, concerned a Wryneck eating a dead bird (presumed to be an Oriental Magpie-Robin *Copsychus saularis*). When disturbed, the Wryneck transferred the carcass to its bill and flew to a higher branch in another tree, where it continued feeding upon it, pinning it under its feet and methodically plucking out the feathers in small clumps in the manner of a raptor. Finally, the Wryneck flew away, still clutching the corpse. It was thought that it had not killed the bird itself but had found the carcass, which was headless, and hence took advantage (Madhav and Victor 2010). Wrynecks are food specialists but, like most woodpeckers, can be opportunistic feeders.

Plant food

Some seeds and berries are occasionally eaten, with elderberries and bilberries among those documented (Harber 1950; Glutz and Bauer 1994). These are, however, merely supplementary items, as Wrynecks have evolved for a mainly insect-based diet. Although they often nest in orchards, using old trees such as apple, apricot, cherry and pear, they rarely eat fruit although it is often plentiful and available, both on the tree and as windfall (pers. obs.). Plant matter found in nest boxes probably arrived there by chance, after being stuck to the viscous tongue when the bird was catching ants (Terhivuo 1976).

Shells

Fragments of shells are sometimes found in Wryneck nests and in the droppings of their chicks. Besides the eggshells of the birds themselves, those of poultry and the shells of marine bivalves and terrestrial snails have also been identified (Terhivuo 1976, 1977, 1983). Adults have been documented as feeding pieces of such shells to their chicks, and both nestlings and parents will also eat their own eggshells (Klaver 1964; Löhrl 1978). In Finland, Terhivuo found that Wrynecks had carried pieces of the shells of both local and exotic marine bivalves and chicken eggs to their nests, which suggested that the birds had visited the nearby coast and poultry farms (exotic bivalves are used in poultry feed). Birds in general will consume shells in order to supplement their diet with calcium, which they need for skeletal growth and egg development. A study in the Netherlands on Great Tits found that snail shells were the main calcium source where eggshell defects did not occur; in areas where the tits exhibited calcium deficiency, snail shells were rarely eaten, the birds instead using other calcium sources such as poultry eggshells and grit (Graveland 1996). It is likely that Wrynecks, too, collect and eat shell fragments for their calcium content.

Non-edible objects

Other items that are clearly non-edible are sometimes discovered in the bottom of nest boxes. Pieces of glass, plastic, wax, porcelain, putty, metal, aluminium foil, grit, fragments of small mammal bones, fish bones and scales, bottle tops, even teeth from a comb and an air-gun pellet have all been recorded (Christensen 1975; Terhivuo 1977; Borgström 1980). Such objects vary in size, shape and colour: white, green, brown, yellow, blue, silver, grey and transparent have been found. In Finland, Terhivuo (1983) revealed that this collecting behaviour was more common than previously believed. Of 121 nest boxes examined around the country, only nine (7.4%) lacked non-edible objects and the average number per nest was 37.4, some nests containing more than 200 objects. Most were found in the post-breeding

FIGURE 14.10 Although they routinely look for food on the ground, Wrynecks will also search on trees and logs. April 2021, Gerecse Hills, Hungary (GG).

waste at the bottom of boxes. With the possible exception of bones, such items have no nutritional value and it was unclear whether any had been fed to the nestlings. Indeed, some would have been too large for even adults to swallow. Nevertheless, chicks do sometimes consume such items, as these have been found in their droppings and in the stomachs of dead birds (Klaver 1964). Christensen (1975) does mention direct feeding and that it resulted in higher levels of nestling mortality than usual. On the other hand, Terhivuo (1983) found that of 64 dead nestlings just seven (11%) had died from eating non-edible objects.

Quite why Wrynecks collect such items is unclear. It has been suggested that grit and small stones aid digestion by helping to break up the hard bodies of ant prey (Klaver 1964; Löhrl 1978). Another idea is that it is linked to searching for alternative food sources when ants are scarce, as for example during bad weather, and the adults are attracted to small glistening objects which they confuse with shiny-bodied ants. Terhivuo (1983) found that this behaviour intensified at the end of the nesting period, during which the parents were most active in feeding their young. The stimuli from such unusual items, together with the strong motivation of the parents to find food for their demanding brood, may have activated a 'picking up' response which resulted in errors in what was collected.

Chapter 15

Relationships

All living things interact with the environment in which they live and with other organisms within it. Interactions among species can be grouped into two types: 'intraspecific' and 'interspecific'. We can define these terms simply as the relationships animals have with others of their own species (intraspecific) and relationships they have with other species (interspecific).

Intraspecific competition

As mentioned in Chapter 10, Wrynecks are not social birds. They generally only engage with each other in the breeding season, when rivals sing and display, and pairs are formed. Conflicts can occur over mates, food resources and particularly nest sites. Indeed, most competition revolves around nest sites as suitable ones are often in high demand. This can even result in the losing pair in a dispute failing to breed (Becker et al. 2014). Physical contact between cospecifics is uncommon and major hostility only really occurs when birds remove or destroy the clutches of others nearby (Menzel 1968; Becker and Tolkmitt 2010). Another type of intraspecific relationship is nest parasitism (see Chapter 12) where an extra-pair female lays her eggs in the nest of another (Fraticelli and Wirz 1991).

Interspecific competition

As discussed in Chapter 13, Wrynecks do not excavate their own nest sites. The list of birds that vie with Wrynecks for cavities, and nest boxes, is long. Examples across Europe include the Pied Flycatcher *Ficedula hypoleuca*, Collared Flycatcher *Ficedula albicollis*, Great Tit, Blue Tit, Coal Tit *Periparus ater*, Marsh Tit, Willow Tit *Poecile montanus*, House Sparrow, Tree Sparrow *Passer*

FIGURE 15.1 Eurasian Nuthatches, like Wrynecks, are unable to excavate their own nest holes and where they coincide the two species compete for cavities. May 2016, Toulon, France (JMB).

montanus, Common Redstart *Phoenicurus phoenicurus*, Eurasian Nuthatch *Sitta europaea*, Common Starling, Hoopoe and even Scops Owl *Otus scops* (Åbro 1962, Menzel 1968, Ruge 1971).

If a suitable cavity is already occupied, sitting birds may be usurped. This usually involves Wrynecks removing any nest material, eggs or nestlings whilst the original occupant is away. Sometimes Wrynecks will sit for hours over several days in front of a cavity where there is a songbird brood, effectively blocking it and preventing the parents from feeding their chicks. On the other hand, if timings allow, already sitting birds may not need to be forced out; that is, they are not usurped but the cavity used once they have finished breeding. Nevertheless, in interspecific encounters over nest sites, Wrynecks are, as we might expect, dominant over some species and submissive to others. They are assertive against smaller birds, and it is not unusual for them to throw out competitors' nestlings, alive or after having pecked them to death (Walzel 1930; Sallay 1984; Hrabovský 2013). Songbirds are aware of this threat and often mob Wrynecks. Although smaller, Great Tits and Tree Sparrows in particular will attempt to fight off an assault on their nests (Matunák 1919). Sometimes they are successful and chase off the larger species (Ruge 1971). As resident birds, tits and sparrows are usually advanced in their breeding cycle, often already incubating or feeding their nestlings when Wrynecks begin to search for a place to nest. In this respect, an observation from the Czech Republic is noteworthy (Hrabec and Damborská 2011). A pair of Great Tits with food in their bills were seen flying in an agitated manner around a nest box which they often used. Upon inspection it was found that a Wryneck was sitting on the tits' brood: three dead nestlings with bloody wounds on their heads and five that were still alive. The disturbed Wryneck left the box and never returned, and the five tit nestlings subsequently fledged successfully.

Wrynecks will even raid active nests in cavities when they are already nesting themselves and have no immediate need of the sites (Ruge 1971; Sallay 1984). They may do this systematically, throwing out the contents of any cavities near

FIGURE 15.2 Though they seldom interact, when it comes to occupying cavities, Hoopoes are dominant over Wrynecks. May 2021, Bologna, Italy (ET).

their own. During a study of Pied Flycatchers, Haartman (1957) observed that many of the cavity nests he was monitoring were often eliminated by Wrynecks without them ultimately using any of the sites. Nevertheless, they often seem to be rather cautious when inspecting holes. This is not unusual, as there is always the possibility of an already sitting occupant becoming aggressive or the site occupied by a larger bird such as a Hoopoe, which in some regions are major rivals for nests and one that is difficult to compete against (Glutz and Bauer 1994). Hoopoes are not, however, competitors for food resources as they mainly feed on Mole Crickets rather than ants (Mermod et al. 2008).

Other wildlife

Various dormice (Gliridae), several arboreal bat species and even insects such as bees, hornets and wasps, also compete for tree hollows, holes and nest boxes. Wrynecks rarely enter into conflicts, nor indeed interact in any way, with these animals (pers. obs.).

FIGURE 15.3 When checking nest boxes care should be taken as European Hornets *Vespa crabro* readily use them and can be hostile to intruders. This queen was inspecting a newly erected Wryneck box in the Vértes Hills, Hungary in May 2021 (GG).

FIGURE 15.4 Two recently laid Wrynecks eggs below an old European Hornet nest. May 2013, Valtice, Czech Republic (JC).

FIGURE 15.5 Green Woodpeckers are also mainly ground foragers with a comparable ant diet to Wrynecks. April 2020, Toulon, France (JMB).

Relationships with other woodpeckers

Wrynecks have a commensal association with several other woodpecker species. They clearly benefit by breeding in the cavities their relatives create, while those species derive no benefit. Nevertheless, woodpeckers sometimes return to the holes they originally excavated and which a Wryneck pair is using. In such cases, Wrynecks can suffer when their clutches and broods are predated, for example by Great Spotted or Syrian Woodpeckers. Wrynecks and Eurasian Green Woodpeckers are sympatric across much of their European breeding ranges, often occupying similar habitats and even preying on the same species of terrestrial ants. Yet competition, and interactions in general between these two species, is surprisingly rare (King and Speight 1974).

Predators and prey

The most important predator versus prey relationship that concerns Wrynecks involves these birds and their main food, ground-dwelling ants (see Chapter 14). The birds themselves are generally not the main prey of any predatory species, avian or mammalian. Nevertheless, as they forage mostly on open ground they are often exposed and thus vulnerable to predators, but probably no more than other species and perhaps less so, as it is likely that their plumage, lacking bright colours, is an advantage in this respect. Even so, Wrynecks are occasionally taken by birds of prey such as hawks and falcons, and mammals like martens and stoats. Locally they may be an important prey resource. A study in Norway found that a surprisingly high number of adults, rather than juveniles, were taken by Sparrowhawks *Accipiter nisus* (Selås 1993). In addition, among birds weighing less than 50 g, the Wryneck was also one of the species most frequently taken by Goshawks *Accipiter gentilis*. This suggests that adults can actually be subject to heavy predation pressure by raptors.

In southern Europe, the Aesculapian Snake *Zamenis longissimus*, a skilled tree climber, often preys on birds' eggs and chicks including those of Wryneck (pers. obs.). Other arboreal snakes, particularly those in the Asiatic breeding range, almost certainly prey on Wrynecks, too, but information is lacking.

In Germany, the Raccoon *Procyon lotor*, an introduced non-native and highly opportunistic species, is known to regularly visit nest boxes to prey on birds and their eggs, including Wrynecks (Becker et al. 2014). Feral and domestic cats are other non-natural predators. Indeed, the problem of countless felines

FIGURE 15.6 Aesculapian Snakes are an occasional threat to nesting Wrynecks. Velebit Mountains, Croatia (GG).

roaming both urban and rural areas is a huge threat to many animals. Their most common targets are common ground-feeding birds such as thrushes, but Wrynecks also fall victim. For example, since 1976 just four foreign-ringed individuals have been found in Britain and of these two were killed by cats. One had travelled 1737 km from Sweden and one 1290 km from Norway (Robinson et al. 2020). A second turned up in Burnham-on-Sea, Somerset, England (where it is an extremely rare visitor) only to be killed by a pet cat. A bird ringed in Kent, England, was also killed by a cat in Finland some ten months later after flying 1823 km, and one which had been ringed as a nestling in Latvia was later re-trapped and released in Hungary only to be taken by a cat in Italy after having flown at least 1891 km (Török 2009).

Chapter 16

Folklore, Mythology and Symbolism

> The sitting bird looks up with jetty eye,
> And waves her head in terror to and fro,
> Speckled and veined with various shades of brown;
> And then a hissing noise assails the clown.
> Quickly, with hasty terror in his breast,
> From the tree's knotty trunk he slides adown
> And thinks the strange bird guards a serpent's nest.
>
> Extract from 'The Wryneck's Nest' by John Clare, 1835.

The English poet John Clare had watched Wrynecks closely. His description, in the verse at the top of this page, of the behaviour of a bird sitting in its nest which 'waves' its head when disturbed clearly illustrates this. No doubt Clare had little trouble in finding these birds in his home county of Northamptonshire in the nineteenth century, as the species was widespread across much of England and Wales back then. Indeed, it was known to William Turner, the 'father of British ornithology', as long ago as 1544 (Bircham 2007). In his autobiography the renowned H. G. Alexander (1974) wrote that, in the early twentieth century, Wrynecks were a 'plentiful breeding species all over south-east England' and that he often saw them (and Red-backed Shrikes) from the train as they perched on fences or wires by Romney Marsh in Kent. Not now. These species have long since gone as breeding birds in rural England.

Like most birds, Wrynecks not only have a natural history but also a human history, a cultural history, and, in England, in that they endure. Throughout the ages, birds have been appreciated, respected and even venerated. They have also been persecuted and demonised. Wrynecks have never attracted people's attention to the extent that eagles, gamebirds and songbirds (to name just a

FIGURE 16.1 A Wryneck poses on a fence-post in Norfolk, England, in April 2011 (NeB). Today the species is only seen in Britain when passing though on migration.

few) have, as they are not huge, nor colourful, nor virtuoso singers, in addition to which they probably do not taste good. Nonetheless, they have not been overlooked entirely. Indeed, they have a long human-connected history. Over 2,500 years ago, in Ancient Babylonia, Wrynecks were considered to be close to the gods. Four gilded images of these birds adorned the Babylonian royal court, evidently placed there to remind rulers to remain humble. In Ancient Greece they were attributed with magical powers which resulted in their being used in a cruel ritual (of which we shall hear more below). In more modern times, these birds have been credited with the ability to invoke good health and their body parts believed to possess medicinal properties. A particularly bizarre example of the latter comes from rural Austria and Bavaria where (in the past, mercifully), gall from a Wryneck was used to thin bushy eyebrows. In more recent times the bird reared its head in popular lore once again when, in the 1990s, soon after the reunification of Germany, the term *Wendehals* was revived and given to people with no political principles, opportunists who conveniently changed their political standpoints, twisting and bending this way and that (Gorman 2017).

Names

Linnaeus gave this species the scientific name *Jynx torquilla*. *Jynx* is derived from the Ancient Greek *iunx*, which was applied both to the bird and to a love charm made with its corpse. *Torquilla* is Renaissance coinage from the Classical Latin *torquere*, meaning to twist or to whirl. In most modern languages the common name refers only to the species' habit of twisting its head awry, rather than the jinx element. In addition, across Europe there is an assortment of folk names that originate from other aspects of its behaviour or from fanciful connotations. Some allude to its arrival in spring, to the weather, to the sowing of crops, to its song, to a perceived relationship with the Cuckoo or to it being an eater of ants. Many refer to its peculiar habit of writhing like a snake. The richness of rural names in English for the Wryneck also tells us something else – just how common this bird must once have been in the British countryside.

Neck-twisting and head-turning

The English vernacular name of this fascinating species can be traced back to the sixteenth century (Bircham 2007). The bird was named 'wry'-neck not because it had a sardonic nature but owing to its habit of contorting and twisting its neck when alarmed (see Chapter 10). This descriptive name is by

FIGURE 16.2 *De draaihals*. From *Onze vogels in huis en tuin* (Our birds in home and garden), 1869–76 by John Gerrard Keuleman. The bird's Dutch name translates as 'turn neck'. WikiCommons.

FIGURE 16.3 *Cuckoo and Wryneck*. From Thomas Pennant's *British Zoology*, Pl. XXXVI. Biodiversity Heritage Library (WikiCommons).

no means exclusive to English; for example, *torcicollo* in Italian, *torcicolo* in Portuguese, *torcecuello* in Spanish, *nyaktekercs* in Hungarian and *krutihlav* in Czech might all be translated as 'neck-twister'. In Welsh it is *gyddfgam* (bent-neck), in Romanian *capîntortura* (head-turner) and in Polish *krętogłów* (twisting head), with older names such as *kręciszyja* (twisting neck) and also *wigłów* (wriggling head). The French name *torcol fourmilier* (neck-twisting ant-eater) is particularly descriptive. In Germanic languages there is the Dutch *draaihals* and German *wendehals* (turn neck) and Danish and Norwegian *vendehals* (turn neck/turn throat).

The theme is present also in many regional languages and dialects. In Spain there are *retorcecuello* from the Zaragoza area, *colltort* in Catalonia (Bernis 1994) and *lepitzuli* in Euskera/Basque (Penas Patiño et al. 2012). Across Italy, *tortacollo, collotorta, torsacol, capitorsolo* and *giracollo* are just some of the local names that nod in the direction of neck- and head-twisting and turning. In Bohemia and Moravia folk names include *krkotoč* (neck-turner), *kroutil* (twister), *kroutílek* (little twister), *vijohlávek* (little head-weaver), *vrtihlav* (head-wagger) and *vrtohlávek* (little head-wagger).

The cuckoo's mate

Across Europe, for reasons that are not obvious, folk names in several languages associate the Wryneck with the totally unrelated Cuckoo. In England, Cuckoo's mate, maid, knave, footman, messenger and marrow ('marrow' meaning 'mate' in the English Midlands) all probably arose owing to the fact that the spring arrival times of these two bird species were roughly the same (when they were both common, we might note). In rural Bohemian there is *Kukačka malá* (little cuckoo). The modern Swedish name *göktyta* is probably derived from *gök-tydaren*, meaning that it arrives 10–12 days before the Cuckoo (*gök*) is heard in spring (Svanberg 2013). The common Finnish name *käenpiika* (Cuckoo's maid) is believed to refer to the Wryneck's habit of diligently cleaning old nest material from holes rather than to any shared spring arrival times; on the same theme there is also a folk name *vanhapiika* (old maid). Another Finnish folk name is *käenelättäjä* (Cuckoo's foster).

Weather and sowing

The Wryneck's song has been interpreted in various ways in relation to the weather in different countries. In Austria, a calling Wryneck signifies coming rain, and common folk names for the species in that country include *giesser*

FIGURE 16.4 The Wryneck was once known as the 'barley bird' in rural England. From *A History of British Birds* by Rev. F. O. Morris (1862). Internet Archive Book Images (WikiCommons).

(pourer) and *gießvogel* (pouring bird). Such names can also refer to other woodpeckers, as they do in other languages. We might also note that the use of such words as *gaiss* and *geiss* may be the result of transliterations of this bird's song. In the Tyrol and Carinthia, the Wryneck is the *regenbitter* (rain begger) and in slang *pipivogel* (pee-pee bird); in Styria, its influence on the weather seemingly changes, as it is the *schniavogel* (snowing bird). In Switzerland, the bird promises good weather, its repetitive song being interpreted as *weib, weib, weib,* prompting country people to declare that winter is over and spring and summer on the way as the bird is calling his wife. A local name from the Asturias region in Spain is *paxarón del agua* (the bird that announces rain), a notion based on the popular belief that Wrynecks (along with some other birds) call more vigorously just before a downpour (Arce and Vázquez 2014). Yet, far to the north in Estonia, the same bird promises dry weather, and just across the Baltic in Finland a rural name *poutahaukka* translates as 'dry-weather hawk'. In southern England, the Wryneck used to be known as the 'barley bird', as it usually appeared just as good weather arrived and that cereal was being sown. Interestingly, most local names for the species in Sweden are associated with the planting of crops in spring, such as *sädeshök* (grain hawk), *sädesgök* (grain cuckoo) and *såsparv* (sow sparrow) (Svanberg 2013).

Singing

Unsurprisingly, the distinctive song of the Wryneck, loudly proclaimed from prominent perches when these birds first arrive back in spring, did not go unnoticed. In Sweden they were heralds, and hearing one was a signal to start sowing wherever one lived, regardless of latitude (Svanberg 2013). In some English counties, such as Surrey, old names such as 'pee bird' or 'pea bird' probably came from its typical *pee-pee-pee* song. In nearby Sussex, 'peel bird' perhaps derived from pee bird (Lockwood 1984). In Spain's Asturias region the most common folk name is *ayayái*, which is again onomatopoeic (Arce and Vázquez 2014). A Hungarian folk name *jajgató madár* (wailing bird) is not onomatopoeic but descriptive.

Ant-eater

The fact that Wrynecks feed primarily on ants did not escape the attention of observant country folk. In Somerset, England, where 'emmet' meant ant, the bird was known as the 'emmet hunter'. In regional languages across

FIGURE 16.5 The 'ant-eater' with a bill full of the insects. June 2012, Vértes Hills, Hungary (AK).

Spain, it was called 'ant-eater', examples including *hormiguero* from Huelva in Andalusia (Bernis 1994), *formiguer* in Valencia (Clavell i Corbera 2002), *formigueru* in Asturias (Arce and Vázquez 2014), *formiguerol* in Menorca, and *formiguer* in Formentera in the Balearic Islands (Clavell i Corbera 2002). In Galicia, this species is the *peto formigueiro* (ant-eating woodpecker) (Penas Patiño et al. 2004). In Italy, too, many dialect names include variations on the word *formiche* (ants), such as in Sardinia *papaformigas*, which means literally 'ant-eater'. An old Czech name *mravenečník* also translates as 'ant-eater' and the official French name *torcol fourmilier* refers to the bird's liking for these insects.

The snakebird

The neck-and-head-twisting behaviour of Wrynecks, which they sometimes perform when threatened or handled, led to their becoming associated with serpents. As long ago as 350 BCE Aristotle wrote of this woodpecker in his *History of Animals*, where 'it has a tongue like those of snakes. For it can stick its tongue out to the breadth of four fingers and pull it back in again. Also,

FIGURE 16.6 The wriggling and writhing movements that Wrynecks often make when handled led to them being called 'snakebirds' by folk across Europe. September 2018, Bourgas, Bulgaria (GG).

it twists its neck round back to front, while the remainder of its body stays motionless, like snakes' (quoted in Ogden 2002).

Aristotle was not alone in making this comparison, as the contortions and sounds which the birds sometimes make when cornered in their nest hole or handled have led many to liken this performance to that of a writhing snake warding off a predator. An old English name is 'snakebird', and in rural Austria folk names include *aderwindl* and *natterwindl* (wriggling adder), *natterfink* (adder finch) and *natterzunge* (adder tongue). In Hungarian, too, the same theme is seen in names such as *kígyómadár* (snakebird) and *sziszegő madár* (hissing bird). Similarly, in rural Czech the Wryneck was the *pěnice hadí* (snake warbler), in Croatian *zmijoglava* (snake head) and in Finnish the *käärmekaula* (snake neck) which, by the way, is the same name as that used by the Finns for the Anhinga! In the Balearic Islands of Spain, the local names *llenguerut* in Mallorca and *llengut* in Ibiza refer to the bird's long tongue (Clavell i Corbera 2002), as do *tiro-lëngo* (stuck-out tongue) an old name from Provence in south-east France. Italian names include *lengua longa* in Liguria, *lingua-longa* in Sicily and *picchio lingua-lunga* in Marche.

The Wryneck's neck- and head-contorting displays are generally considered to be examples of behavioural mimicry (see Chapter 10). The function of this extraordinary performance remains unclear, and it is therefore not surprising that in the past a mythology and lore arose around it and the birds – regrettably, not always of a positive nature. Predictably, as 'snakebirds', Wrynecks consequently became associated with witchcraft and evil in general. Like serpents, they have been demonised and accused of wielding malevolent power. They were, for instance, believed to be sinister and cunning, luring ants into their bill by lying on the ground and playing dead (Tate 2007). In reality, of course, Wrynecks lick up ants with their extended tongue while very much alive.

Wryneck wheels

In Ancient Greece, Wrynecks were known as the *lynx* or *lunx* bird and were believed to possess mystical powers. A whole mythology developed around the birds, especially in regard to love, desire and eroticism, as they were said to become possessed during courtship, excitedly stretching and twisting their necks and rotating their heads. The bird was known also as the *kinaidios*, which implied being promiscuous. Sophocles (*c.*497–405 BCE) called it the 'vigilant guardian of love'. Poets recounted how spurned women would attempt to harness this power in order to charm and entice back errant lovers, using a macabre ritual to do so.

A captured Wryneck was impaled, wings spread, upon a small wheel which was spun around until it made droning and humming sounds as incantations or spells were chanted. These 'jynx wheels' or 'wryneck wheels' were Hellenic love charms that date back to at least the eighth century BCE (Karageorghis 1989). They were made of wood or metal and were often ornate with two eyes in the centre through which two threads were passed. The alternate pulling of the threads and the twisting of the wheel produced a whistling sound and this may have been an attempt to mimic the distinctive song of the bird. Illustrations of such wheels on vases, sometimes in the hands of Eros, survive in museums and private collections. The *Suda*, a Byzantine Greek encyclopaedia from the tenth century CE, describes the *iynx* as 'a bird suited to the evils of love-charms'. Aphrodite, the goddess of love, was said to have used the combined power of the bird, a finely gilded wheel and potions based on olive oil to compel the sorceress Medea to fall in love with Jason. In the fifth century BCE, Pindar wrote in a Pythian ode that Aphrodite 'yoking the dappled wryneck all asprawl to the inescapable wheel, brought down from Olympus the bird of madness for the first time' (Tate 2007). In his poem 'The Sorceress', Theocritus (*c.*300–260 BCE) describes how Simaetha performed spells and rituals to entice back her lost lover Delphis, repeatedly chanting 'Wryneck, draw this man to my house' (Ogden 2002).

Although no wryneck wheels are mentioned, the bird appears elsewhere in Greek mythology. Philyra, a shape-shifting goddess associated with healing, divination, beauty and perfume, was aligned with the Wryneck, and the bird was also an icon for a college of priestesses who served Thetis. In yet another tale, the nymph Iynx had her voice taken away by Hera, queen of the gods, who gave it to another nymph, Echo. Iynx sought revenge by casting a spell to make Hera's husband Zeus fall in love with the priestess Io. When Hera learned what Iynx had done, she cursed her, turning her into a Wryneck. In some interpretations, Iynx is said to have invented the wryneck wheel and used it when casting her spell.

All in all, it is unclear why the Wryneck was so strongly associated with passion, eroticism and the wheel charm in Ancient Greece. It may be the case that the bird's name came first and the use of the bird in the wheel came to mean 'charm' or 'spell'. Indeed, the connection between the birds and the whole erotic mythos is something of a mystery. Scholars have so far failed to provide a satisfactory explanation. One proposal is that the Greek symbolism of the Wryneck as a sensual animal arose from the legend of Hera and Iynx. Another is that the bird was chosen for the wheel because it uttered snake-like hisses and sometimes darted its long serpent-like tongue from its mouth.

Perhaps it was simply that the writhing and twisting that these birds perform were interpreted as an erotic mating dance.

It was also believed that if a man wore a lapis lazuli ring on which Aphrodite had been engraved and that enclosed a Wryneck's right eye he would have sexual favours bestowed upon him. In addition, this would also help him to influence other men and ensure his success in any legal action brought against him. Apparently, the left eye did the same for women (Ogden 2002). Mercifully, such hocus-pocus is no longer practised in Greece, nor anywhere else, which is just as well since Wrynecks have many other threats and hazards to contend with. Curiously, despite a plethora of folk names in languages right across Europe, and the immense influence that Ancient and Classical Greece had on the European cultures that followed, none refers to the erotic notions that the Greeks associated with Wrynecks.

An adult Wryneck clings to the side of a tree. May 2021, Novo Yankovo, Bulgaria (DG).

References

Åbro, A. (1962) Vendehalsens Reir. *Sterna* 5: 146–7.

Adamian, M. S. and Klem, D. (1999) *Handbook of the Birds of Armenia*. American University of Armenia, Oakland, California.

Alderson, H. (1897) Wonderful egg-producing powers of the Wryneck. *Zoologist* 4: 511–12.

Alexander, H. G. (1974) *Seventy Years of Birdwatching*. Poyser, Berkhamsted.

Allen, J. A. (2004) Avian and mammalian predators of terrestrial gastropods. In Barker, G. M. (ed), *Natural Enemies of Terrestrial Molluscs*. CABI Publishing, Wallingford.

Alter, H. (2002) Ungewöhnlicher Nistplatz vom Wendehals (*Jynx torquilla*) bei Cottbus. *Otis – Zeitschrift für Ornithologie und Avifaunistik in Brandenburg und Berlin* 10: 158–60.

Andersen, E. S. (2018) Forbedret vurdering af ynglebestande af Vendehals. *Dansk Ornitologisk Forenings Tidsskrift* 112: 19–28. https://doi.org/10.22439/dansoc.v28i1.5599

Arce, L. M. and Vázquez, V. M. (2014) *Aves de Asturias. Guía de identificación*. Ediciones Mundi-Prensa, Madrid.

Ash, J. and Atkins, J. (2009) *Birds of Ethiopia and Eritrea: An Atlas of Distribution*. Helm, London.

Assandri G. et al. (2018) A matter of pipes: Wryneck *Jynx torquilla* habitat selection and breeding performance in an intensive agroecosystem. *Journal of Ornithology* 159: 103–14. https://doi.org/10.1007/s10336-017-1479-y

Ayé, R., Schweizer, M. and Roth, T. (2012) *Birds of Central Asia*. Helm, London.

Axelsson, C., Nömm, M., Carlsson, H. and Carlsson, L. (1997) Projekt göktyta *Jynx torquilla*: biotopval och häckningsframgång (Project Wryneck *Jynx torquilla*: habitat selection and breeding success). *Ornis Svecica* 7: 35–7. (In Swedish with English summary.)

Bailey, A. M. (1945) Wryneck from Cape Prince of Wales, Alaska. *Auk* 64: 456. https://doi.org/10.2307/4080394

Bairlein, F. (1988) How do migratory songbirds cross the Sahara? *Trends in Ecology and Evolution* 3: 191–4. https://doi.org/10.1016/0169-5347(88)90005-5

Bairlein, F., Dierschke, V., Salewski, V., Geiter, O., Hüppop, K., Köppen, U. and Fiedler, W. (eds) (2014) *Atlas des Vogelzugs: Ringfunde deutscher Brut- und Gastvögel* (Bird Migration Atlas: Ring recoveries of German breeding and visiting birds). Frankfurt: AULA-Verlag. 335–6. (In German with English summary.)

Baker, J. (2016) *Identification of European Non-Passerines*. British Trust for Ornithology, Thetford. pp. 420–2.

Battisti, C. and Dodaro, G. (2016) Mapping bird assemblages in a Mediterranean urban park: evidence for a shift in dominance towards medium-large body sized species after 26 years. *Belgian Journal of Zoolology* 146: 81–9. https://doi.org/10.26496/bjz.2016.43

Bautz, C. (1998) Vegetationskomplexe der Eberstädter Streuobstwiesen unter besonderer Berücksichtigung der Habitatansprüche des Wendehalses. *Collurio (Zeitschrift für Vogel und Naturschutz in Südhessen)* 16: 29–35.

Becker, D. and Tolkmitt, D. (2007) Zur Brutbiologie des Wendehalses *Jynx torquilla* im nordöstlichen Harzvorland – Die Gelegegröße (Breeding biology of the Wryneck *Jynx torquilla* at the north-eastern border of the Harz mountains – the clutch size). *Ornithologische Jahresberichte des Museum Heineanum* 25: 29–47. (In German with English summary.)

Becker, D. and Tolkmitt, D. (2010) Zur Brutbiologie des Wendehalses *Jynx torquilla* im nordöstlichen Harzvorland III. Schlupferfolg (Breeding biology of the Wryneck *Jynx torquilla* at the northeastern border of the Harz mountains III. Hatching success). *Ornithologische Jahresberichte des Museum Heineanum* 28: 1–14. (In German with English summary.)

Becker, D., Tolkmitt, D. and Nicolai, B. (2013) Individualerkennung anhand der Steuerfedern beim Wendehals *Jynx torquilla* (Individual recognition by tail feathers of Wryneck *Jynx torquilla*). *Ornithologische Jahresberichte des Museum Heineanum* 31: 21–31. (In German with English summary.)

Becker, D., Tolkmitt, D. and Nicolai, B. (2014) Zur Brutbiologie des Wendehalses *Jynx torquilla* im nordöstlichen Harzvorland: IV. Brutgröße und Fortpflanzungsziffer (Breeding biology of the Wryneck *Jynx torquilla* at the north-eastern border of the Harz mountains: IV. Breeding success). *Ornithologische Jahresberichte des Museum Heineanum* 32: 43–57. (In German with English summary.)

Benabbas-Sahki, I. et al. (2015) Trophic availabilities impact in prey-ants on food ethology of Eurasian Wryneck *Jynx torquilla mauretanica* (Linnaeus 1758) (Piciformes: Picidae) in eastern Plain of Mitidja, Algeria. *Egyptian Journal of Biological Pest Control* 25: 709–17.

Benz, B. W., Robbins, M. B. and Peterson, A. T. (2006) Evolutionary history of woodpeckers and allies (Aves: Picidae): placing key taxa on the phylogenetic tree. *Molecular Phylogenetics and Evolution* 40: 389–99. https://doi.org/10.1016/j.ympev.2006.02.021

Bergier, P., Thévenot, M. and Qninba, A. (2017) *Oiseaux du Sahara Atlantique Marocain (Birds of the Moroccan Atlantic Sahara)*. SEOF Edicions, Paris.

Bernis, F. (1994) *Diccionario de nombres vernáculos de aves*. Editorial Gredos, Madrid.

Bijlsma, R. G. (2014) Broed- en foerageergedrag van Draaihalzen *Jynx torquilla* (Breeding and foraging behaviour of Wrynecks *Jynx torquilla*). *Drentse Vogels* 28: 78–100. (In Dutch with English summary.)

Bijlsma, R. G. (2016) The tongue of the Wryneck *Jynx torquilla* in a wider perspective. *Drentse Vogels* 30: 50–4.

Bircham, P. (2007) *A History of Ornithology*. Collins, London.

BirdLife International (2020) Species factsheet: *Jynx ruficollis*. http://www.birdlife.org (Accessed 29/11/2020).

BirdLife International (2021) Species factsheet: *Jynx torquilla*. http://www.birdlife.org (Accessed 04/01/2021).

Bitz, A. and Rohe, W. (1993) Nahrungsökologische Untersuchungen am Wendehals (*Jynx torquilla*) in Rheinland-Pfalz. Beih. Veröff. *Naturschutz Landschaftspflege Baden-Württemberg* 67: 83–100.

Blasco-Zumeta, J. and Heinze, G. M. (2017) Atlas de Identificacíon de las Aves de Aragón. 287 Wryneck. Ibercaja Aula en Red. http://listafuglestasjon.no/pdf/Jtorquilla.pdf

Bock, W. J. (2015) Evolutionary morphology of the woodpeckers (Picidae). In Winkler, H. and Gusenleitner, F. (eds), *Developments in Woodpecker Biology*. Biologiezentrum des Oberösterreichischen Landesmuseums, Linz, Austria.

Booth, J., Booth, C. J. and Pattenden, B. (1968) Wryneck apparently wintering in Cornwall. *British Birds* 61: 174.

Borgström, E. (1980) Göktyta drar in främmande föremål i sitt bo (Wryneck, *Jynx torquilla*, brings strange objects into its nest). *Vår Fågelvärld* 39: 101–2. (In Swedish with English summary.)

Borrow, N. and Demey, R. (2011) *Birds of Senegal and The Gambia*. Helm, London.

Bozó, L. and Csörgő, T. (2020) Changes in spring arrival dates of central European bird species over the past 100 years. *Acta Zoologica Academiae Scientiarum Hungaricae* 66: 283–98. https://doi.org/10.17109/AZH.66.3.283.2020

Brejcha, J. (2019) Viper as a Batesian model – its role in an ecological community. *Biosemiotics* 12: 25–38. https://doi.org/10.1007/s12304-019-09347-x

Brichetti, P. and Fracasso, G. (2020) *The Birds of Italy. Vol. 2: Pteroclidae–Locustellidae*. Edizioni Belvedere, Latina. 121–4.

Brünner, K. and Rödl, T. (2018) Erfolgreiche Bestandsstützung beim Wendehals *Jynx torquilla* in den ostmittelfränkischen Sanden (Successful protection of dwindling Wryneck *Jynx torquilla* populations in eastern Middle Franconia). *Ornithologische Anzeiger* 57: 45–51. (In German with English summary.)

Burton, H., Evans, T. L. and Weir, D. N. (1970) Wrynecks breeding in Scotland. *Scottish Birds* 6: 154–6.

Busche, G. (2004) Zum Durchzug des Wendehalses (*Jynx torquilla*) an der Deutschen Bucht (Helgoland und schleswig-holsteinische Küste) 1965–1998 (Passage migration of wrynecks (*Jynx torquilla*) at the German Bight (Helgoland and Schleswig-Holstein) in the period 1965–1998). *Vogelwarte* 42: 344–51. (In German with English summary.)

Bussmann, J. (1941) Beitrag zur Kenntnis der Brutbiologie des Wendehalser (*Jynx t. torquilla*). *Schweizerisches Archiv für Ornithologie* 1.5: 467–80.

Carbonell, R. (2012) Torcecuello euroasiático *Jynx torquilla*. In: SEO/BirdLife. 2012. *Atlas de las aves en invierno en España 2007–2010*: 350–1. Ministerio de Agricultura, Alimentación y Medio Ambiente/SEO/BirdLife, Madrid.

Cave, F. O. and Macdonald, J. D. (1955) *Birds of The Sudan*. Oliver and Boyd, Edinburgh/London.

Chantler, P. J. (1991) Call of migrant Wryneck. *British Birds* 84: 195.

Chernel, I. (1903) Kesztyűben fészkelő nyaktekercs. *Aquila* 1–4: 254–5.

Chikara, O. (2019) *Birds of Japan*. Lynx Edicions, Barcelona.

Chmielewski, S. (2019) The importance of old, traditionally managed orchards for breeding birds in the agricultural landscape. *Polish Journal of Environmental Studies* 28: 1–8. https://doi.org/10.15244/pjoes/94813

Chodkiewicz, T. et al. (2015) Ocena liczebności populacji ptaków lęgowych w Polsce w latach 2008–2012 (Population estimates of breeding birds in Poland in 2008–2012). *Ornis Polonica* 56: 149–89. (In Polish with English abstract.)

Chowdhury, S. U. (2020) Birds of the Bangladesh Sundarbans: status, threats and conservation recommendations. *Forktail* 36: 35–46.

Christensen, J. H. (1975) Vendehalsen som ynglefugl i Vestjylland. *Danske Fugle* 24: 201–3.

Clavell i Corbera, J. (2002) *Catàleg dels ocells dels Països Catalans (Catalunya, País Valencià, Illes Balears, Catalunya Nord)*. Lynx Edicions, Barcelona.

Clements, J. F., Schulenberg, T. S., Iliff, M. J., Billerman, S. M., Fredericks, T. A., Sullivan, B. L. and Wood, C. L. (2019) The eBird/Clements Checklist of Birds of the World: v2019. Cornell. https://www.birds.cornell.edu/clementschecklist/download/ (Accessed 10/11/2020).

Collette, J. (2008) Les oiseaux du verger en Normandie. *Le Cormoran* 16: 31–57.

Contejean, G. (1998) Elevage particulier d'un jeune Torcol fourmilier *Jynx torquilla* (Peculiar rearing circumstances of a Wryneck *Jynx torquilla* nestling). *Nos Oiseaux* 45: 250. (In French with English summary.)

Coudrain, V., Arlettaz, R. and Schaub, M. (2010) Food or nesting place? Identifying factors limiting Wryneck populations. *Journal of Ornitholology* 151: 867–80. https://doi.org/10.1007/s10336-010-0525-9

Cramp, S. (ed.) (1985) *Handbook of the Birds of Europe, the Middle East and North Africa. The Birds of the Western Palearctic*. Vol. 4. Oxford University Press, Oxford.

del Hoyo, J. and Collar, N. J. (2014) *HBW and BirdLife International Illustrated Checklist of the Birds of the World*. Vol. 1. Lynx Edicions, Barcelona.

del Val, E., Dreiser, C., Finkbeiner, W. and Förschler, M. (2018) Der Wendehals *Jynx torquilla* als Brutvogel der Windwurfflächen im Nordschwarzwald (Der Wendehals *Jynx torquilla* als Brutvogel der Windwurfflächen im Nordschwarzwald). *Vogelwarte* 56: 9–13. (In German with English summary.)

Demongin, L. (2016) *Identification Guide to Birds in the Hand*. Beauregard-Vendon. p. 207.

Desfayes, M. (1969) A possible hybrid *Jynx ruficollis × torquilla*. *Bulletin of the British Ornithologists' Club* 89: 110–12.

Dobay, L. (1927) *Der Brut parasitimus des Kuckuchs, mit Desonderer Rücksicht auf die siebenbürgischen Verhältnisse*. Xth. International Congress of Zoology, Budapest. 855.

DOF (2020) https://dofbasen.dk (Database of BirdLife Denmark).

Dolenec, Z. (2020) Contribution to the oology of the Eurasian wryneck *Jynx torquilla*. *Natura Croatica* 29: 123–8. https://doi.org/10.20302/NC.2020.29.11

Donovan, J. W. (1968) Wryneck in Pembrokeshire in winter. *British Birds* 61: 173–4.

Dowsett, R. J. (1968) Migrants at Malam'fatori, Lake Chad, Spring 1968. *Bulletin of the Nigerian Ornithologists' Society* 5: 53–6.

Dowsett, R. J. (1969) Migrants at Malam'fatori, Lake Chad, Autumn 1968. *Bulletin of the Nigerian Ornithologists' Society* 6: 39–45.

Dowsett-Lemaire, F. and Dowsett, R. J. (2014) *The Birds of Ghana: An Atlas and Handbook*. Tauraco Press, Liège.

Dowsett-Lemaire, F. and Dowsett, R. J. (2019) *The Birds of Benin and Togo: An Atlas and Handbook*. Tauraco Press, Sumène.

Du, C., Liu, L. I., Liu, Y. and Fu, Z. (2020) The complete mitochondrial genome of the Eurasian wryneck *Jynx torquilla* (Aves: Piciformes: Picidae) and its phylogenetic inference. *Zootaxa*: 4810 (2). https://doi.org/10.11646/zootaxa.4810.2.8

Dunning, J. B. Jr., Beheler, A., Crowder, M., Andrews, S. and Weiss, R. (2002) A Eurasian Wryneck specimen from southern Indiana. *North American Birds* 56: 264–7.

Dutour, M. et al. (2019) Hissing like a snake: bird hisses are similar to snake hisses and prompt similar anxiety behavior in a mammalian model. *Behavioral Ecology and Sociobiology* 74: 1. https://doi.org/10.1007/s00265-019-2778-5

Dvorak, M. (2019) Österreichischer Bericht gemäß. Artikel 12 der Vogelschutzrichtlinie, EG – Berichtszeitraum 2013 bis 2018. BirdLife Österreich, Wien.

Dymond, J. N. (1991) *The Birds of Fair Isle*. Privately published by the author.

Eason, P., Rabia, B. and Attum, O. (2015) Hunting of migratory birds in North Sinai, Egypt. *Bird Conservation International*. https://doi.org/10.1017/S0959270915000180

Eaton, M. et al. (2015) Birds of Conservation Concern 4: the population status of birds in the UK, Channel Islands and Isle of Man. *British Birds* 108: 708–46.

Eaton, M. and Holling, M. (2020) Rare Breeding Birds in the UK in 2018. *British Birds* 113.

EBCC (2020) Pan-European Common Bird Monitoring Scheme. European Bird Census Council. https://pecbms.info/trends-and-indicators/species-trends/species/jynx-torquilla/?search=jynx

Eck, S. and Geidel, B. (1974) Die flügel-schwanz-verhaeltnisse paläarktischer Wendehälse (*Jynx torquilla*) (Aves, Picidae). (The wing-tail of Palearctic Wrynecks *Jynx torquilla*). *Zoologische Abhandlungen Staatliches Museum fur Tierkunde Dresden* 32: 257–65. (In German with English summary.)

Elts, J., Leito, A., Leivits M., Luigujõe L., Nellis R., Ots M., Tammekänd, I. and Väli, Ü. (2019) Eesti lindude staatus, pesitsusaegne ja talvine arvukus 2013–2017 (Status and numbers of Estonian birds, 2013–2017). *Hirundo* 32: 1–39. (In Estonian with English summary.)

Eriksen, J. and Porter, R. (2018) *Birds of Oman*. Helm, London.

Forejt, M. and Syrbe, R-U. (2019) The current status of orchard meadows in Central Europe: multi-source area estimation in Saxony (Germany) and the Czech Republic. *Moravian Geographical Reports* 27: 217–28. https://doi.org/10.2478/mgr-2019-0017

Forrester, R. and Andrews, I. (2007) *The Birds of Scotland*. Vol. 2: 956–9. Scottish Ornithologist's Club, Aberlady.

Fransson, T., Jansson L., Kolehmainen T., Kroon, C. and Wenninger, T. (2017) EURING list of longevity records for European birds. https://euring.org/data-and-codes/longevity-list (Accessed 10/03/2021).

Fraser, P. A. and Rogers, M. J. (2006) Report on scarce migrant birds in Britain in 2003. Part 1: American Wigeon to Wryneck *British Birds* 99: 74–91.

Fraticelli, F. and Wirz, A. (1991) Evidence of intraspecific nest parasitism in Wryneck *Jynx torquilla*. *Avocetta* 15: 65.

Freitag, A. (1996) Le régime alimentaire du Torcol fourmilier (*Jynx torquilla*) en Valais (Suisse). *Nos Oiseaux* 43: 497–512.

Freitag, A. (2000) La photographie des nourrissages: une technique originale d'étude du régime alimentaire des jeunes Torcols fourmiliers *Jynx torquilla*. *Alauda* 68: 81–93. (In French with English summary.)

Freitag, A., Martinoli, D. and Urzelai, J. (2001) Monitoring the feeding activity of the nesting birds with an autonomous system: case study of the endangered Wryneck *Jynx torquilla*. *Bird Study* 48: 102–9. https://doi.org/10.1080/00063650109461207

Garcías, P. (2010) Revisió de l'estatus del formiguer *Jynx torquilla* a Mallorca. *Anuari Ornitologic de les Balears* 25: 29–41.

Gatter, W. (1997) *Birds of Liberia*. Pica Press, Mountfield.

Gedeon, K. et al. (2014) *Atlas Deutscher Brutvogelarten* (Atlas of German Breeding Birds). Stiftung Vogelmonitoring Deutschland und Dachverband Deutscher Avifaunisten, Münster. 364–5. (In German with English summary.)

Geiser, S., Arlettaz, R. and Schaub, M. (2008) Impact of weather variation on feeding behaviour, nestling growth and brood survival in wrynecks *Jynx torquilla*. *Journal of Ornithology* 149: 597–606. https://doi.org/10.1007/s10336-008-0305-y

Gill, F., Donsker, D. and Rasmussen, P. (eds) (2020) IOC World Bird List. https://doi.org/10.14344/IOC.ML.11.1

Ginn, H. B. and Melville, D. S. (1983) *Moult in Birds*. Guide 19. British Trust for Ornithology, Tring.

Glutz von Blotzheim, U. N. and Bauer, K. M. (eds) (1994) *Handbuch der Vögel Mitteleuropas. Band 9. Columbiformes-Piciformes*. AULA-Verlag Gmbh, Wiesbaden.

González, J. V., Gómez, J. R. and Muñoz, B. (2002) *El torcecuello. Determinación de la edad y el sexo, reproducción y fenología en el Noroeste peninsular ibérico*. Sociedad Asturiana de Historia Natural y Ecología.

Goodge, W. R. (1972) Anatomical evidence for phylogenetic relationships among woodpeckers. *The Auk* 89: 65–85. https://doi.org/10.2307/4084060

Gorman, G. (2004) *Woodpeckers of Europe: A Study of the European Picidae*. Bruce Coleman, Chalfont St Peter.

Gorman, G. (2014) *Woodpeckers of the World: The Complete Guide*. Helm, London.

Gorman, G. (2017) *Woodpecker*. Reaktion, London.

Gorman, G. (2020) Reverse mounting by three European *Dendrocopos* woodpeckers. *British Birds* 113: 180–2.

Gorman, G. (2021) A review of snake mimicry in the Eurasian Wryneck. *British Birds* 114: 480–4.

Gorman, G. et al. (2021) Nyaktekercs – *Jynx torquilla* – Eurasian Wryneck. In Szép, T. et al. (eds.), *Magyarország madáratlasza – Bird Atlas of Hungary*. Agrárminisztérium, Magyar Madártani és Természetvédelmi Egyesület, Budapest.

Graveland, J. (1996) Avian eggshell formation in calcium-rich and calcium-poor habitats: the importance of snail shells and anthropogenic calcium sources. *Canadian Journal of Zoology* 74: 1035–44. https://doi.org/10.1139/z96-115

Gregory, R. D. et al. (2007) Population trends of widespread woodland birds in Europe. *Ibis* 49: 78–97. https://doi.org/10.1111/j.1474-919X.2007.00698.x

Grimmett, R., Roberts, T. and Inskipp, T. (2008) *Birds of Pakistan*. Helm, London.

Grimmett, R., Inskipp, C. and Inskipp, T. (2011) *Birds of the Indian Subcontinent*. Helm, London.

Grimmett, R., Inskipp, C., Inskipp, T. and Sherub (2019) *Birds of Bhutan and the Eastern Himalayas*. Helm, London.

Gustafsson, R., Hjort, C., Ottosson, U. and Hall, P. (2021) *Birds at Lake Chad and in the Sahel of NE Nigeria 1997–2000*. The Lake Chad Bird Migration Project.

Haartman, L. von. (1957) Adaptation in hole-nesting birds. *Evolution* 11: 339–47. https://doi.org/10.1111/j.1558-5646.1957.tb02902.x

Hansen, W. and Synnatzschke, J. (2015) *Die Steuerfedern der Vögel Mitteleuropas* (The Tail Feathers of the Birds of Central Europe). World Feather Atlas Foundation. (In German with English summaries.)

Harber, D. D. (1950) Wryneck feeding on Elderberries. *British Birds* 43: 310.

Harrison, C. (1974) *A Field Guide to the Nests, Eggs and Nestlings of European Birds with North Africa and the Middle East*. Collins, London.

Hedenström, A. and Lindström, A. (1990) High body masses of migrating Wrynecks *Jynx torquilla* in southern Sweden. *Vogelwarte* 35: 165–8.

Herrando, S., Brotons, L., Estrada, J., Cuallar, S. and Anton, M. (eds) (2011) *Atles dels ocells de Catalunya a l'hivern 2006–2009*. ICO/Lynx Edicions. Barcelona.

Hirschfeld, E. (1995) *Birds in Bahrain: a study of their migration patterns 1990–1992*. Hobby Publications, Dubai.

Höfling, E. and Alvarenga, H. M. F. (2001) Osteology of the Shoulder Girdle in the Piciformes, Passeriformes and Related Groups of Birds. *Zoologischer Anzeiger – Journal of Comparative Zoology* 240: 196–208. https://doi.org/10.1078/0044-5231-00016

Hogstad, O. (2006) Reproductive differences between woodpeckers and secondary hole-nesters. *Ornis Norvegica* 29: 110–23. https://doi.org/10.15845/on.v29i0.183

Holling, M. et al. (2012) Rare breeding birds in the United Kingdom in 2010. *British Birds* 105: 397–8.

Holloway, S. (1996) *The Historical Atlas of Breeding Birds of Britain and Ireland: 1875–1900*. Poyser, London.

Hrabec, J. and Damborská, M. (2011) Neobvyklé zjištění krutihlava obecného (*Jynx torquilla*) na hnízdě s mláďaty sýkory koňadry (*Parus major*) (Unusual record of the Eurasian Wryneck on the nest with Great Tit chicks). *Acrocephalus* 26: 115. (In Czech with English summary.)

Hrabovský, M. (2013) Rekordně početná snůška krutihlava obecného (*Jynx torquilla*) hnízdícího v budce obsazené původně sýkorou koňadrou (*Parus major*). (A record clutch of the Eurasian Wryneck (*Jynx torquilla*) breeding in a nest-box occupied originally by a Great Tit (*Parus major*). *Crex* 32: 124–8. (In Czech with English summary.)

Iankov, P. (ed.) (2007) *Atlas of Breeding Birds in Bulgaria*. Conservation Series, Book 10: 356–7. Bulgarian Society for the Protection of Birds, Sofia.

Iovchenko, N. P. and Kovalev, V. A. (2005) Иовченко Н. П., Ковалев В. А. Постювенальная линька вертишейки *Jynx torquilla* L. в юго-восточном Приладожье Post juvenile moult in the Wryneck *Jynx torquilla* L. in the south-eastern Ladoga area. In Iovchenko, N. P. (ed.), *Ornithological Studies in the Ladoga Area*. 158–72. St Petersburg University Press. (In Russian with English summary.)

Iovchenko, N. P. and Kovalev, V. A. (2016) Wryneck *Jynx torquilla*. In Noskov, G. A., Rymkevich, T. A. and Gaginskaya, A. R. (eds) (*Migration of Birds of Northwest Russia. Non-passerines*). ANO LA, St Petersburg. 558–61.

Isenmann, P. and Moali, A. (2000) *Oiseaux d'Algérie* (Birds of Algeria). SEOF Edicions, Paris.

Isenmann, P., Gaultier, H., El Hili, A., Azafzaf, H., Dlensi, H. and Smart, M. (2005). *Oiseaux de Tunisie* (Birds of Tunisia). SEOF Edicions, Paris.

Isenmann, P., Benmergui, M., Browne, P., Ba, A. D., Diagana, C. H., Diawara, Y. and El Abidine Ould Sidaty, Z. (2010) *Oiseaux de Mauritanie* (Birds of Mauritania). SEOF Edicions, Paris.

Issa, N. and Muller, Y. (eds) (2015) *Atlas des Oiseaux de France métropolitaine. Nidification et Présence hivernale*. Delachaux et Niestlé, Paris. Vol. 2: 796–9.

Jacob, J-P. et al. (2010) *Atlas des Oiseaux Nicheurs de Wallonie 2001–2007*. Aves-Natagora & Département de l'Etude du milieu Naturel et Agricole, Serie Faune-Flore-Habitats n° 5. AVES a.s.b.l. Gembloux.

Jenni, L. and Winkler, R. (2020) *The Biology of Moult in Birds*. Helm, London.

Judson, O. P. and Bennett, A. T. D. (1992) 'Anting' as food preparation: formic acid is worse on an empty stomach. *Behavioral Ecology and Sociobiology* 31: 437–9. https://doi.org/10.1007/BF00170611

Kajtoch, Ł. (2017) The importance of traditional orchards for breeding birds: the preliminary study on central European example. *Acta Oecologica* 78: 53–60. https://doi.org/10.1016/j.actao.2016.12.010

Karageorghis, V. (1989) Two votive lynx-wheels from Cyprus. In *Architecture et poésie dans le monde grec. Hommage à Georges Roux*: 263–8. Maison de l'Orient et de la Méditerranée Jean Pouilloux, Lyon.

Kasambe, R., Dudhe, N., Wagh, G., Kale, M. and More, K. (2014) Bird biodiversity in agricultural landscape in Vidarbha, Maharashtra. *Newsletter for Birdwatchers* 54: 64–8.

Keller, V. et al. (2020). *European Breeding Bird Atlas 2: Distribution, Abundance and Change*. European Bird Census Council and Lynx Edicions, Barcelona. 512–13.

Kennedy, A. S. (2019) First record of Eurasian Wryneck *Jynx torquilla* for Tanzania. *Scopus* 39: 43.

Ķerus, V., Dekants, A., Auniņš, A. and Mārdega, L. (2021) *Latvijas Ligzdojošo Putnu Atlanti 1980–2017* (Latvian Breeding Bird Atlas 1980–2017). Latvian Ornithological Society, Rīga.

Kervyn, T. and Xhardez, C. (2006) Utilisation de l'espace par le Torcol fourmilier (*Jynx torquilla*) en Ardenne lors d'une nidification uniparentale (Uniparental nesting care in Wryneck (*Jynx torquilla*). *Aves* 43: 65–72. (In French with English summary.)

Kessler, J. (2014) Fossil and subfossil bird remains and faunas from the Carpathian Basin. *Ornis Hungarica* 22: 65–125. https://doi.org/10.2478/orhu-2014-0019

Kessler, J. (2016) Picidae in the European fossil, subfossil and recent bird faunas and their osteological characteristics. *Ornis Hungarica* 24: 96–114. https://doi.org/10.1515/orhu-2016-0006

Kiat, Y. and Izhaki, I. (2017) Non-moulted primary coverts correlate with rapid primary moulting. *Journal of Avian Biology* 48: 380–6. https://doi.org/10.1111/jav.00939

King, B. and Speight, M. C. D. (1974) Anting-like behaviour and food of Wryneck. *British Birds* 67: 388–9.

Kirby, V. C. (1980) An adaptive modification in the ribs of woodpeckers and piculets (Picidae). *The Auk* 97: 521–32.

Kirwan, G. M., Martins, R. P., Eken, G. and Davidson, P. (1999) A checklist of the birds of Turkey. *Sandgrouse* 21 (Suppl. 1) 1–32.

Klaver, A. (1964) Waarnemingen over de biologie van de Raaihals *Jynx torquilla* L. *Limosa* 37: 221–31.

Knaus, P., Sattler, T., Schmid, H., Strebel, N. and Volet, B. (2020) *The State of Birds in Switzerland: Report 2020*. Swiss Ornithological Institute, Sempach.

Krams, I. et al. (2014) Hissing calls improve survival in incubating female great tits *Parus major*. *Acta Ethologica* 17: 83–8. https://doi.org/10.1007/s10211-013-0163-3

Krištín A., Demko M. and Pačenovský, S. (2014) *Červený zoznam vtákov Slovenska* (Red List of Birds in Slovakia). SOS/BirdLife Slovensko.

Laesser, J. and van Wijk, R. E. (2017) Postponed moult of primary-coverts untangles the ageing of Wrynecks *Jynx torquilla*. *Ringing and Migration* 32: 87–103. https://doi.org/10.1080/03078698.2017.1437889

Langslow, D. R. (1977) Weight increases and behaviour of Wrynecks on the Isle of May. *Scottish Birds* 9 (5): 262–7.

Languy, M. (2019) *The Birds of Cameroon: Their status and distribution*. Series 'Studies in Afrotropical Zoology', Vol. 299. Tervuren: Royal Museum for Central Africa.

Lehman, P. E. (2005) Fall bird migration at Gambell, St. Lawrence Island, Alaska. *Western Birds* 36: 30.

Leiber, A. (1907) Vergleichende Anatomie der Spechtzunge. *Zoologica* 6: 51.

Linkola, P. (1978) Häckningsbiologiska undersökningar av göktyta i Finland 1952–1977 (On the breeding biology of the wryneck *Jynx torquilla* in Finland). *Anser*, Suppl. 3: 155–162. Proceedings of the first Nordic Congress of Ornithology 1977.

Lockwood, W. B. (1984) *The Oxford Book of British Bird Names*. Oxford University Press, Oxford.

Löhrl, H. (1978) Beiträge zur Ethologie und Gewichtsentwicklung beim Wendehals *Jynx torquilla* (Notes on ethology and weight development in the Wryneck *Jynx torquilla*). *Ornithologische Beobachter* 75: 193–201. (In German with English summary.)

Lovegrove, R., Williams, G. and Williams, I. (1994) *Birds in Wales*. Poyser, London.

MacKinnon, J. and Phillipps, K. (2000) *A Field Guide to the Birds of China*. Oxford University Press, Oxford.

Madhav, N. V. and Victor, J. R. (2010) Wryneck *Jynx torquilla* feeding on bird in Sundarbans, West Bengal, India. *Indian Birds* 7: 48.

Makatsch, W. (1955) *Der Brutparasitismus in der Vogelwelt*. Neumann Verlag, Radebeul, Berlin.

Martí, R. and Del Moral, J. C. (eds) (2003) *Atlas de las Aves Reproductoras de España*. Dirección General de Conservación de la Naturaleza-Sociedad Española de Ornitología, Madrid.

Martín, A. and Lorenzo, J. A. (2001) *Aves del archipiélago canario*. Francisco Lemus Editor, La Laguna, Tenerife.

Martínez, J. and Domínguez-Santaella, M. (1997) Distribución y caracterización del hábitat del Torcecuello (*Jynx torquilla*) en la provincia de Málaga (Distribution and Habitat characterisation of the Wrynecks (*Jynx torquilla*) in Malaga). *Actas de las XII Jornadas Ornitológicas Españolas* 84: 39–45. (In Spanish with English summary.)

Matunák, M. (1919) Széncinege és nyaktekercs harca. *Aquila* 26: 121.

Menzel, H. (1968) *Der Wendehals (Jynx torquilla)*. Die Neue Brehm-Bücherei, Band 392. A. Ziemsen Verlag. Wittenberg-Lutherstadt.

Mermod, M., Reichlin, T. S., Schaub, M. and Arlettaz, R. (2008) Wiedehopfpaar zieht Wendehalsnestlinge bis zum Ausfliegen auf (Eurasian Hoopoe *Upupa epops* pair raises Eurasian Wryneck *Jynx torquilla* nestlings until they fledge). *Ornithologische Beobachter* 105: 153–160. (In German with English summary.)

Mermod, M., Reichlin, T. S., Arlettaz, R. and Schaub, M. (2009) The importance of ant-rich habitats for the persistence of the Wryneck *Jynx torquilla* on farmland. *Ibis* 151: 731–42. https://doi.org/10.1111/j.1474-919X.2009.00956.x

Miettinen, J. (2002) Age determination in woodpeckers. In Pechacek, P. and d'Oleire-Oltmanns, W. (eds), *International Woodpecker Symposium*: 127–31. Forschungsbericht 48, Nationalparkverwaltung Berchtesgaden.

Mihelič, T., Kmecl, P., Denac, K., Koce, U., Vrezec, A. and Denac, D. (eds) (2019) *Atlas ptic Slovenije. Popis gnezdilk 2002–2017* (Slovenian Breeding Bird Atlas). DOPPS, Ljubljana. 248–9. (In Slovenian with English summary.)

Mikusiński, G. and Angelstam, P. (1997) European woodpeckers and anthropogenic habitat change: a review. *Vogelwelt* 118: 277–83.

Mischenko, A. L. (ed.) (2017) Estimation of numbers and trends for birds of the European Russia (European Red List of birds. Moscow, Russian Society for Bird Conservation and Study.

Mlíkovský, J. (2006) Egg size in birds of southern Bohemia: an analysis of Rudolf Prazny's collection. *Sylvia* 42: 112–16.

MNS Bird Conservation Council (2015) *A Checklist of the birds of Malaysia*. Conservation Publication No. 14. Malaysian Nature Society, Kuala Lumpur.

Møller, A. P., Flensted-Jensen, E. and Liang, W. (2021) Behavioral snake mimicry in breeding tits. *Current Zoology* 67: 27–33. https://doi.org/10.1093/cz/zoaa028

Monk, J. F. (1963) The past and present status of the Wryneck in the British Isles. *Bird Study* 10: 112–32. https://doi.org/10.1080/00063656309476045

Morel, G. J. and Morel, M. Y. (1990) *Les Oiseaux de Sénégambie*. ORSTOM, Paris.

Morozov, N. S. (2015) Why do birds practice anting? *Biology Bulletin Reviews* 5: 353–65. https://doi.org/10.1134/S2079086415040076

Mulhauser, B. and Zimmermann, J-L. (2014) Croissance des oisillons de Torcol fourmilier *Jynx torquilla*, de l'éclosion à l'envol (Growth of Wryneck *Jynx torquilla* chicks from hatching to fledging). *Nos Oiseaux* 61: 181–9. (In French with English summary.)

Murdoch, D. A. and Betton, K. F. (2008) A checklist of the birds of Syria. *Sandgrouse* Suppl. 2.

Nenadović, Ž. (2008) Uzastopno gnežđenje vijoglave *Jynx torquilla*, bele pliske *Motacilla alba* i velike senice *Parus major* u istoj šupljini (The successive nesting of Eurasian Wryneck *Jynx torquilla*, White Wagtail *Motacilla alba* and Great Tit *Parus major* in the same hole). *Ciconia* 17: 100. (In Serbian with English summary.)

Ogden, D. (2002) *Magic, Witchcraft, and Ghosts in the Greek and Roman Worlds: A Sourcebook*. Oxford University Press, Oxford.

Orsini, P. (1997) L'hivernage du Torcol fourmilier *Jynx torquilla* en France continentale (Wintering of Wryneck in continental France). *Ornithos* 4: 21–7. (In French with English summary.)

Pakkala, T., Tiainen, J., Piha, M. and Kouki, J. (2019) Hole life: survival patterns and reuse of cavities made by the Lesser Spotted Woodpecker *Dendrocopos minor*. *Ardea* 107: 173–81. https://doi.org/10.5253/arde.v107i2.a4

Palm, V. et al. (2009) The spring timing of arrival of migratory birds: dependence on climate variables and migration route. *Ornis Fennica* 86: 97–108.

Peal, R. E. F. (1968) The distribution of the Wryneck in the British Isles 1964–1966. *Bird Study* 15: 111–26. https://doi.org/10.1080/00063656809476191

Peal, R. E. F. (1973) Studies of less familiar birds: Wryneck. *British Birds* 66: 66–72.

Pearson, D. and Jackson, C. (2016) Review of Kenya bird records 2011–2014. *Scopus* 36: 31.

Pearson, D. J. and Turner, D. A. (1986) The less common Palearctic migrant birds of Uganda. *Scopus* 10: 61–82.

Penas Patiño, X. M., Pedreira López, C. and Silvar, C. (2004) *Guía das aves de Galicia.* Baía Edicions, A Coruña.

Penas Patiño, X. M., Pedreira López, C. and Silvar, C. (2012) *Guía de aves de Euskal Herría.* Sua Edizioak, Bilbao.

Petrovici, M. et al. (2015) *Atlasul pasarilor de interes comumitar din Romania.* ROS and Milvus Group, Romania. 312–13.

Pinilla, J. (2021) Modelling the winter distribution of the Eurasian Wryneck (*Jynx torquilla*) in Iberia. *Airo* 28: 3–12.

Price, J (2017) The potential impacts of climate change on the biodiversity of Norfolk. *Transactions of the Norfolk and Norwich Naturalists' Society* 50: 1–8.

Puhlmann, E. (1914) Zur Brutgeschichte der Wendehalse. *Ornithologische Monatsschrift* 39: 205–7.

Puzović, S. et al. (2015) *Birds of Serbia: Breeding Population Estimates and Trends for the Period 2008–2013*. Bird Protection and Study Society of Serbia and Department of Biology and Ecology, University of Novi Sad.

Ram, D., Lindström, Å., Pettersson, L. B. and Caplat, P. (2020) Forest clear-cuts as habitat for farmland birds and butterflies. *Forest Ecology and Management* 473: 118239. https://doi.org/10.1016/j.foreco.2020.118239

Ramadan-Jaradi, G., Itani, F., Hogg, J., Serhal, A. and Ramadan-Jaradi, M. (2020) Updated checklist of the birds of Lebanon, with notes on four new breeding species in spring. *Sandgrouse* 42: 186–238.

Rasmussen, P. C. and Anderton, J. C. (2005) *Birds of South Asia: The Ripley Guide.* Vols 1 and 2. Smithsonian Institution and Lynx Edicions, Washington, D. C. and Barcelona.

Redman, N., Stevenson, T. and Fanshawe, J. (2011) *Birds of the Horn of Africa: Ethiopia, Eritrea, Djibouti, Somalia, and Socotra.* Helm, London.

Reichlin, T. S., Schaub, M., Menz, M. H. M. et al. (2009) Migration patterns of Hoopoe *Upupa epops* and Wryneck *Jynx torquilla*: an analysis of European ring recoveries. *Journal of Ornithology* 150: 393–400. https://doi.org/10.1007/s10336-008-0361-3

Reiter, A. (2013) Neobvyklé umístění hnízd mlynaříka dlouhoocasého (*Aegithalos caudatus*) a krutihlava obecného (*Jynx torquilla*) v Národním parku Podyjí (An unusual nest location of the Long-tailed Tit and the Eurasian Wryneck in the Podyjí National Park.) *Crex* 32: 129–34. (In Czech with English summary.)

Richardson, C. (1990) *The Birds of the United Arab Emirates.* Hobby Publications, Dubai.

Richardson, C. and Porter, R. (2020) *Birds of Cyprus.* Helm, London.

Robinson, R. A., Leech, D. I. and Clark, J. A. (2020) The Online Demography Report: Bird ringing and nest recording in Britain and Ireland in 2019. British Trust for Ornithology, Thetford (http://www.bto.org/ringing-report, created on 10 September 2020).

Robson, C. (2011) *A Field Guide to the Birds of South-East Asia*. New Holland, London.

Rodwell, S. P., Sauvage, A., Rumsey, S. J. R. and Bräunlich, A. (1996) An annotated check-list of birds occurring at the Parc National des Oiseaux du Djoudj in Senegal, 1984–1994. *Malimbus* 18: 74–111.

Ruge, K. (1971) Beobachtungen am Wendehals *Jynx torquilla*. *Ornithologische Beobachter* 68: 9–33.

Ruge, K., Bastian, H-V. and Bruland, W. (1988) *Der Wendehals*. Vogelkunde Bücherei 5. Verlag Opus data, Rottenburg.

Ryttman, H. (2003) Breeding success of Wryneck *Jynx torquilla* during the last 40 years in Sweden. *Ornis Svecica* 13: 25–8.

Sallay, Z. (1984) Nyaktekercs (*Jynx torquilla*) fészekpusztítása. *Madártani tájékoztató*, April–June: 90–1.

Salvan, J. (1967–9) Contribution à l'étude des oiseaux du Tchad. *L'Oiseau et R.F.O.* 37: 255–84; 38: 53–85, 127–50, 249–73; 39: 38–69.

Sanderson, F. J., Donald, P. F., Pain, D. J. et al. (2006) Long-term population declines in Afro-Palearctic migrant birds. *Biological Conservation* 131: 93–105. https://doi.org/10.1016/j.biocon.2006.02.008

Sauvage, A., Rumsey, S. and Rodwell, S. (1998) Recurrence of Palearctic birds in the lower Senegal river valley. *Malimbus* 20: 33–53.

Scebba, S. and Lövei, G. L. (1985) Winter recurrence, weights and wing lengths of wrynecks *Jynx torquilla* on a southern Italian Island. *Ringing and Migration* 6: 83–6. https://doi.org/10.1080/03078698.1985.9673861

Schaub, M., Martinez, N., Tagmann-Ioset, A., Weisshaupt, N. et al. (2010) Patches of bare ground as a staple commodity for declining ground-foraging insectivorous farmland birds. *PLoS ONE* 5 (10): e13115. https://doi.org/10.1371/journal.pone.0013115

Schaub, M., Reichlin, T. S., Abadi, F., Kéry, M., Jenni, L. and Arlettaz, R. (2012) The demographic drivers of local population dynamics in two rare migratory birds. *Oecologia* 168: 97–108. https://doi.org/10.1007/s00442-011-2070-5

Schneider, W. (1961) Tambourinage du Torcol. *Nos Oiseaux* 26: 50.

Schulze, M. (2008) Der Wendehals *Jynx torquilla* als Erdhöhlenbrüter (The Wryneck *Jynx torquilla* as a burrowing bird). *Ornithologische Jahresberichte des Museum Heineanum* 26: 109–116. (In German with English summary.)

Sehhatisabet, M. E. et al. (2006) Further significant extensions of migrant distribution and breeding and wintering ranges in Iran for over sixty species. *Sandgrouse* 28: 146–55.

Selås, V. (1993) Selection of avian prey by breeding Sparrowhawks *Accipiter nisus* in southern Norway: The importance of size and foraging behaviour of prey. *Ornis Fennica* 70: 144–54.

Sermet, E. (2001) Un nid de torcol fourmilier *Jynx torquilla* a double ouverture (A Wryneck nesting cavity with two entrances). *Nos Oiseaux* 484: 254–5 (In French with English summary.)

Shimmings, P. and Øien, I. J. (2015) *Bestandsestimater for norske hekkefugler.* NOF (BirdLife Norway) rapport 2: 152–3.

Shirihai, H., Dovrat, E. and Christie, D. (1996) *The Birds of Israel: A Complete Avifauna and Bird Atlas of Israel.* Academic Press, London.

Short, L. L. (1982) *Woodpeckers of the World.* Delaware Museum of Natural History, Greenville, Delaware.

Short, L. L. (1988a) *Jynx torquilla.* In Fry, C. H., Keith, S. and Urban, E. K. (eds), *The Birds of Africa.* Vol. 3. Academic Press, San Diego. 514–15.

Short, L. L. (1988b) *Jynx ruficollis.* In Fry, C. H., Keith, S. and Urban, E. K. (eds), *The Birds of Africa.* Vol. 3. Academic Press, San Diego. 515–17.

Short, L. L. and Bock, W. J. (1972) Possible hybrid *Jynx* is an aberrant *Jynx ruficollis. Bulletin of the British Ornithologists' Club* 92: 28–91.

Sibley, C. G. (1955) Behavioral mimicry in the Titmice (Paridae) and certain other birds. *Wilson Bulletin* 67: 128–32.

Sibley, C. G. and Ahlquist, J. E. (1990) *Phylogeny and Classification of Birds: A Study in Molecular Evolution.* Yale University Press, New Haven and London.

Singer, D. S., Benson, T. A. McCaskie, G. and Stahl, J. (2020) The 43rd Annual Report of the California Bird Records Committee: 2017 Records. *Western Birds* 51: 2–26. https://doi.org/10.21199/WB51.1.1

Solti, B. (2010) A Mátra Múzeum madártani gyûjteménye III. Németh Márton tojásgyűjtemény. *Folia Historico-naturalia Musei Matraensis, Supplementum* 5: 1–275.

Sorace, A. and Gustin, M. (2010) Bird species of conservation concern along urban gradients in Italy. *Biodiversity Conservation* 19: 205–1. https://doi.org/10.1007/s10531-009-9716-1

Sovon Dutch Centre for Field Ornithology (2021) https://www.sovon.nl/nl/soort/8480. (Accessed 04/01/2021).

Spina, F. and Volponi, S. (2008) *Atlante della Migrazione degli Uccelli in Italia. 1. non-Passeriformi* (Italian Bird Migration Atlas. Vol. 1 non-Passeriformes). Ministero dell'Ambiente e della Tutela del Territorio e del Mare, Istituto Superiore per la Protezione e la Ricerca Ambientale (ISPRA). Tipografia CSR-Roma. 782–6. (In Italian with English summary.)

Šťastný, K. and Bejček, V. (2021) *Atlas hnízdního rozšíření ptáků v České republice 2014–2017* (Atlas of Breeding Birds in the Czech Republic). Aventinum, Praha.

Steinfatt, O. (1941) Beobachtungen über das Leben des Wendehalses *Jynx t. torquilla. Beitr. Fortpflanzungsbiol. Vögel* 17: 185–200.

Stêpniewska, K., El-Hallah, A. and Busse, P. (2011) Migration dynamics and directional preferences of passerine migrants in Azraq (E Jordan) in spring 2008. *Ring* 33: 1–2: 3–25. https://doi.org/10.2478/v10050-011-0001-9

Stevenson, T. and Fanshawe, J. (2002) *Birds of East Africa: Kenya, Tanzania, Uganda, Rwanda, Burundi.* Poyser, London.

Stoddard, M. C. (2012) Mimicry and masquerade from the avian visual perspective. *Current Zoology* 58: 630–48. https://doi.org/10.1093/czoolo/58.4.630

Stone, R. C. (1954) 'Anting' by Wryneck. *British Birds* 47: 312.

Stresemann, E. and Stresemann, V. (1966) *Die Mauser der Vögel.* Friedländer, Berlin.

Sutter, E. (1941) Beitrag zur Kenntnis der postembryonalen Entwicklung des Wendehalses (*Jynx t. torquilla* L.). *Schweizerisches Archiv für Ornithologie* 1: 481–508.

Svanberg. I. (2013) *Fåglar i svensk folklig tradition.* Dialogos Förlag, Stockholm.

Svensson, S. (2000) *Övervakning av fåglars populationsutveckling. Årsrapport för 1999.* Ekologiska institutionen, Lunds universitet, Lund.

Tarboton, W. R. (1976) Aspects of the biology of *Jynx ruficollis. Ostrich* 47: 99–112. https://doi.org/10.1080/00306525.1976.9639545

Tarboton, W. R. (2005) Red-throated Wryneck *Jynx ruficollis.* In Hockey, P. A. R., Dean, W. R. J. and Ryan, P. G. (eds), *Roberts – Birds of Southern Africa.* 7th edn. Trustees of the John Voelcker Bird Book Fund, Cape Town.

Tate, P. (2007) *Flights of Fancy. Birds in Myth, Legend and Superstition.* Random House, London.

Taylor, S. (1993) Wryneck *Jynx torquilla.* In Gibbons, D. W., Reid, J. B. and Chapman, R. A. (eds), *The New Atlas of Breeding Birds in Britain and Ireland: 1988–1991.* Poyser, London. 262–3.

Terhivuo, J. (1976) Terrestrial snails in the diet of the Wryneck *Jynx torquilla. Ornis Fennica* 53: 47.

Terhivuo, J. (1977) Occurrence of strange objects in nests of the Wryneck *Jynx torquilla. Ornis Fennica* 54: 66–72.

Terhivuo, J. (1983) Why does the Wryneck *Jynx torquilla* bring strange items to the nest? *Ornis Fennica* 60: 51–7.

Teufelbauer, N. and Seaman, B. (2021) *Monitoring der Brutvögel Österreichs – Bericht über die Saison 2020.* BirdLife Österreich, Wien.

Tohidifar, M. (2008) Review of the current status of the Eurasian Wryneck *Jynx torquilla,* Eurasian Treecreeper *Certhia familiaris* and Wallcreeper *Tichodroma muraria* in Iran. *Podoces* 3: 97–131.

Tolkmitt, D., Becker, D. and Kormann, Urs. (2020) Zum Einfluss von Legebeginn und Bruttyp auf die Rekrutierungswahrscheinlichkeit beim Wendehals *Jynx torquilla* (Influence of the onset of egg laying and the brood type on the recruiting probability in Wrynecks *Jynx torquilla*). *Ornithologische Jahresberichte des Museum Heineanum* 35: 103–15. (In German with English summary.)

Török, J. (2009) Nyaktekercs/Wryneck *Jynx torquilla.* In Csörgő, T. et al. (eds), *Magyar madárvonulási atlasz* (Hungarian Bird Migration Atlas). Kossuth kiadó, Budapest. 384–5. (In Hungarian with English summary.)

Turner, K. and Gorman, G. (2021) The instrumental signals of the Eurasian Wryneck (*Jynx torquilla*). *Ornis Hungarica* 29: 98–107. https://doi.org/10.2478/orhu-2021-0007

Vaisanen, R. (2001) Neljän maalinnun pesimäkantojen romahdukset (Steep recent decline in Finnish populations of Wryneck, Wheatear, Chiffchaff and Ortolan Bunting.) *Linnut* 36: 14–15. (In Finnish with English summary.)

Valkama, J., Vepsäläinen, V. and Lehikoinen, A. (2011). *Suomen III lintuatlas.* Luonnontieteellinen keskusmuseo ja ympäristöministeriö (*The Third Finnish Breeding Bird Atlas*) http://atlas3.lintuatlas.fi (Accessed 12/12/2020).

van den Berk, V. (1990) The rapid movement of a Turkish-ringed Wryneck to Beirut, Lebanon. *OSME Bulletin* 24: 15–17.

van Duivendijk, N. (2011) *Advanced bird ID handbook: The western Palearctic.* New Holland, London.

van Wijk, R. E. and Tizón, M. F. (2016) Wintering habitat selection by Eurasian Wrynecks *Jynx torquilla* in the west of the Iberian Peninsula. *Ardeola* 63: 349–356. https://doi.org/10.13157/arla.63.2.2016.sc1

van Wijk, R. E., Schaub, M., Tolkmitt, D., Becker, D. and Hahn, S. (2013) Short-distance migration of Wrynecks *Jynx torquilla* from central European populations. *Ibis* 155: 886–90. https://doi.org/10.1111/ibi.12083

Varga, F. (1981) Újabb odúkészítő nyaktekercs (*Jynx torquilla*) Zagyvarónán. *Madártani tájékoztató*, July–Sept.: 177–8.

Vaurie, C. (1959) Systematic notes on Palearctic birds. No. 37. Picidae: The Subfamilies Jynginae and Picumninae. *American Museum Novitates*, No. 1963.

Vermeersch, G. et al. (2020). *Broedvogels in Vlaanderen 2013–2018. Recente status en trends van in Vlaanderen broedende vogelsoorten.* Mededelingen van het Insti tuut voor Natuur en Bosonderzoek, Brussel.

Vickery, J. A. et al. (2014) The decline of Afro-Palaearctic migrants and an assessment of potential causes. *Ibis* 156: 1–22. https://doi.org/10.1111/ibi.12118

Villiers, A. (1950) Contribution à l'étude de l'Aïr – Oiseaux. *Bulletin de l'Institut français d'Afrique noire* 12: 345–85.

Vogel, R. (1997) Wryneck *Jynx torquilla*. In Hagemeijer, E. J. M. and Blair, M. J. (eds), *The EBCC Atlas of European Breeding Birds: Their distribution and abundance*. Poyser, London.

Vogrin, M. (2011) Overlooked traditional orchards: their importance for breeding birds. *Studia Universitatis Babes-Bolyai, Biologia* 56: 3–9.

Walzel, J. (1930) A kék cinege fiókáit pusztító nyaktekercs. *Aquila* 36/37: 317.

Warga, K. (1926) Harminchármat tojó nyaktekercs. *Aquila* 32–33: 263.

Wassink, A. (2015) *The New Birds of Kazakhstan*. Arend Wassink, Texel, The Netherlands.

Weisshaupt, N., Arlettaz, R., Reichlin, T. S., Tagmann-Ioset, A. and Schaub, M. (2011): Habitat selection by foraging Wrynecks *Jynx torquilla* during the breeding season: identifying the optimal habitat profile. *Bird Study* 58: 111–19. https://doi.org/10.1080/00063657.2011.556183

Wernham, C., Toms, M., Marchant, J., Clark, J., Siriwardena, G. and Baillie, S. (2002) *The Migration Atlas: Movements of the birds of Britain and Ireland*. Poyser, London.

Wink, M. et al. (2011) Mating system, paternity and sex allocation in Eurasian Wrynecks (*Jynx torquilla*). *Journal of Ornithology* 152: 983–9. https://doi.org/10.1007/s10336-011-0684-3

Winkler, H. (2015) Phylogeny, biogeography and systematics. In Winkler, H. and Gusenleitner, F. (eds), *Developments in Woodpecker Biology*. Biologiezentrum des Oberösterreichischen Landesmuseums, Linz, Austria.

Winkler, H. and Christie, D. A. (2002) Family Picidae (Woodpeckers) In: del Hoyo, J., Elliot, A. and Sargatal, J. (eds), *Handbook of the Birds of the World*. Vol. 7. Lynx Edicions, Barcelona.

Winkler, H. and Short, L. (1978) A comparative analysis of acoustical signals in pied woodpeckers (Aves, Picoides). *Bulletin of the American Muswum of Natural History* 160: 1–109.

Winkler, H., Christie, D. A. and Nurney, D. (1995) *Woodpeckers: A Guide to the Woodpeckers, Piculets and Wrynecks of the World*. Pica Press, Robertsbridge, UK.

Winkler, H., Gamauf, A., Nittinger, F. and Haring, E. (2014) Relationships of Old World woodpeckers (Aves: Picidae): new insights and taxonomic implications. *Annalen des Naturhistorischen Museums Wien, Serie B (Botanik und Zoologie)* 116: 69–86.

Winkler, H., Christie, D. A. and Kirwan, G. M. (2020) Eurasian Wryneck (*Jynx torquilla*), version 1.0. In del Hoyo, J., Elliott, A., Sargatal, J., Christie, D. A. and de Juana, E. (eds), *Birds of the World*. Cornell Lab of Ornithology, Ithaca, New York. https://doi.org/10.2173/bow.eurwry.01

Winkler, R. (2013) *Mauserumfang und Altersbestimmung von Spechten*. Ornithologisches Informationsblatt. Swiss Ornithological Institute, Sempach.

Wirdheim, A. (ed.) (2020) *Sveriges fåglar 2020*. BirdLife Sverige.

Witherby, H. F., Jourdain, F. C. R., Ticehurst, N. F. and Tucker, B. W. (1938) *The Handbook of British Birds*, Vol. 2. London.

Yosef, R. and Markovets, M. (2009) Spring bird migration phenology in Eilat, Israel. *ZooKeys* 31: 193–210. https://doi.org/10.3897/zookeys.31.107

Yosef, R. and Zduniak, P. (2011) Migration and staging patterns of the Wryneck (*Jynx torquilla*) at Eilat, Israel. *Israel Journal of Ecology and Evolution* 57: 247–56. https://doi.org/10.1560/IJEE.57.3.247

Yoshimura, M., Hirata, T., Nakajima, A. and Onoyama, K. (2003) Ants found in scats and pellets taken from the nests of the Japanese Wryneck *Jynx torquilla japonica*. *Ornitholgical Science* 2 (2): 127–31. https://doi.org/10.2326/osj.2.127

Young, H. G., Tonge, S. J. and Wilson, D. (1993) Wryneck on passage roosting in reeds. *British Birds* 86: 20.

Zabka, H. (1980) Zur funktionellen Bedeutung der Instrumentallaute europäischer Spechte unter besonderer Berücksichtigung von *Dendrocopos major* und *Dendrocopos minor* (The functional importance of the instrumental sounds of European woodpeckers with special reference to *Dendrocopos major* and *Dendrocopos minor*). *Mitteilungen aus dem Zoologischen Museum in Berlin*, 56 Supplementheft, Annalen für Ornithologie 4: 51–76 (In German with English summary.)

Zimmerman, D. A., Turner, D. A. and Pearson, D. J. (1996) *The Birds of Kenya and Northern Tanzania*. Helm, London.

Zingg, S., Arlettaz, R. and Schaub, M. (2010) Nestbox design influences territory occupancy and reproduction in a declining, secondary cavity-breeding bird. *Ardea* 98: 67–75. https://doi.org/10.5253/078.098.0109

Zmihorski, M., Hebda, G., Eggers, S., Mansson, J., Abrahamsson, T., Czeszczewik, D., Walankiewicz, W. and Mikusiñski, G. (2019) Early post-fire bird community in European boreal forest: comparing salvage-logged with non-intervention areas. *Global Ecology and Conservation* 18: e00636. https://doi.org/10.1016/j.gecco.2019.e00636

Zwarts, L., Bijlsma, R. G., van den Kamp, J. and Wymenga, E. (2009) *Living on the Edge: Wetlands and birds in a changing Sahel*. KNNV Publishing, Zeist.

Index

Page numbers in *italics* indicate images.